LINEAR AND NONLINEAR CONJUGATE GRADIENT-RELATED METHODS

SIAM PROCEEDINGS SERIES LIST

Neustadt, L. W., Proceedings of the First International Congress on Programming and Control (1966)

Hull, T. E., Studies in Optimization (1970)

Day, R. H. and Robinson, S. M., Mathematical Topics in Economic Theory and Computation (1972)

Proschan, F. and Serfling, R. J., Reliability and Biometry: Statistical Analysis of Lifelength (1974)

Barlow, R. E., Reliability & Fault Tree Analysis: Theoretical & Applied Aspects of System Reliability & Safety Assessment (1975)

Fussell, J. B. and Burdick, G. R., Nuclear Systems Reliability Engineering and Risk Assessment (1977)

Erisman, A. M., Neves, K. W., and Dwarakanath, M. H., Electric Power Problems: The Mathematical Challenge (1981)

Bednar, J. B., Redner, R., Robinson, E., and Weglein, A., Conference on Inverse Scattering: Theory and Application (1983)

Santosa, Fadil, Symes, William W., Pao, Yih-Hsing, and Holland, Charles, Inverse Problems of Acoustic and Elastic Waves (1984)

Gross, Kenneth I., Mathematical Methods in Energy Research (1984)

Babuska, I., Chandra, J., and Flaherty, J., Adaptive Computational Methods for Partial Differential Equations (1984)

Boggs, Paul T., Byrd, Richard H., and Schnabel, Robert B., Numerical Optimization 1984 (1985)

Angrand, F., Dervieux, A., Desideri, J. A., and Glowinski, R., Numerical Methods for Euler Equations of Fluid Dynamics (1985)

Wouk, Arthur, New Computing Environments: Parallel, Vector and Systolic (1986)

Fitzgibbon, William E., Mathematical and Computational Methods in Seismic Exploration and Reservoir Modeling (1986)

Drew, Donald A. and Flaherty, Joseph E., Mathematics Applied to Fluid Mechanics and Stability: Proceedings of a Conference Dedicated to R.C. DiPrima (1986)

Heath, Michael T., Hypercube Multiprocessors 1986 (1986)

Papanicolaou, George, Advances in Multiphase Flow and Related Problems (1987)

Wouk, Arthur, New Computing Environments: Microcomputers in Large-Scale Computing (1987)

Chandra, Jagdish and Srivastav, Ram, Constitutive Models of Deformation (1987)

Heath, Michael T., Hypercube Multiprocessors 1987 (1987)

Glowinski, R., Golub, G. H., Meurant, G. A., and Periaux, J., First International Conference on Domain Decomposition Methods for Partial Differential Equations (1988)

Salam, Fathi M. A. and Levi, Mark L., Dynamical Systems Approaches to Nonlinear Problems in Systems and Circuits (1988)

Datta, B., Johnson, C., Kaashoek, M., Plemmons, R., and Sontag, E., Linear Algebra in Signals, Systems and Control (1988)

Ringeisen, Richard D. and Roberts, Fred S., Applications of Discrete Mathematics (1988)

McKenna, James and Temam, Roger, ICIAM '87: Proceedings of the First International Conference on Industrial and Applied Mathematics (1988)

Rodrigue, Garry, Parallel Processing for Scientific Computing (1989)

Caflish, Russel E., Mathematical Aspects of Vortex Dynamics (1989)

Wouk, Arthur, Parallel Processing and Medium-Scale Multiprocessors (1989)

Flaherty, Joseph E., Paslow, Pamela J., Shephard, Mark S., and Vasilakis, John D., Adaptive Methods for Partial Differential Equations (1989)

Kohn, Robert V. and Milton, Graeme W., Random Media and Composites (1989)

Mandel, Jan, McCormick, S. F., Dendy, J. E., Jr., Farhat, Charbel, Lonsdale, Guy, Parter, Seymour V., Ruge, John W., and Stüben, Klaus, Proceedings of the Fourth Copper Mountain Conference on Multigrid Methods (1989)

Colton, David, Ewing, Richard, and Rundell, William, Inverse Problems in Partial Differential Equations (1990)

Chan, Tony F., Glowinski, Roland, Periaux, Jacques, and Widlund, Olof B., Third International Symposium on Domain Decomposition Methods for Partial Differential Equations (1990)

Dongarra, Jack, Messina, Paul, Sorensen, Danny C., and Voigt, Robert G., Proceedings of the Fourth SIAM Conference on Parallel Processing for Scientific Computing (1990)

Glowinski, Roland and Lichnewsky, Alain, Computing Methods in Applied Sciences and Engineering (1990)

Coleman, Thomas F. and Li, Yuying, Large-Scale Numerical Optimization (1990)

Aggarwal, Alok, Borodin, Allan, Gabow, Harold, N., Galil, Zvi, Karp, Richard M., Kleitman, Daniel J., Odlyzko, Andrew M., Pulleyblank, William R., Tardos, Éva, and Vishkin, Uzi, Proceedings of the Second Annual ACM-SIAM Symposium on Discrete Algorithms (1990)

Cohen, Gary, Halpern, Laurence, and Joly, Patrick, Mathematical and Numerical Aspects of Wave Propagation Phenomena (1991)

Gómez, S., Hennart, J. P., and Tapia, R. A., Advances in Numerical Partial Differential Equations and Optimization: Proceedings of the Fifth Mexico-United States Workshop (1991)

Glowinski, Roland, Kuznetsov, Yuri A., Meurant, Gérard, Périaux, Jacques, and Widlund, Olof B., Fourth International Symposium on Domain Decomposition Methods for Partial Differential Equations (1991)

Alavi, Y., Chung, F. R. K., Graham, R. L., and Hsu, D. F., Graph Theory, Combinatorics, Algorithms, and Applications (1991)

Wu, Julian J., Ting, T. C. T., and Barnett, David M., Modern Theory of Anisotropic Elasticity and Applications (1991)

Shearer, Michael, Viscous Profiles and Numerical Methods for Shock Waves (1991)

Griewank, Andreas and Corliss, George F., Automatic Differentiation of Algorithms: Theory, Implementation, and Application (1991)

Frederickson, Greg, Graham, Ron, Hochbaum, Dorit S., Johnson, Ellis, Kosaraju, S. Rao, Luby, Michael, Megiddo, Nimrod, Schieber, Baruch, Vaidya, Pravin, and Yao, Frances, Proceedings of the Third Annual ACM-SIAM Symposium on Discrete Algorithms (1992)

Field, David A. and Komkov, Vadim, Theoretical Aspects of Industrial Design (1992)

Field, David A. and Komkov, Vadim, Geometric Aspects of Industrial Design (1992)

Bednar, J. Bee, Lines, L. R., Stolt, R. H., and Weglein, A. B., Geophysical Inversion (1992)

O'Malley, Robert E. Jr., ICIAM 91: Proceedings of the Second International Conference on Industrial and Applied Mathematics (1992)

Keyes, David E., Chan, Tony F., Meurant, Gérard, Scroggs, Jeffrey S., and Voigt, Robert G., Fifth International Symposium on Domain Decomposition Methods for Partial Differential Equations (1992)

Dongarra, Jack, Messina, Paul, Kennedy, Ken, Sorensen, Danny C., and Voigt, Robert G., Proceedings of the Fifth SIAM Conference on Parallel Processing for Scientific Computing (1992)

Corones, James P., Kristensson, Gerhard, Nelson, Paul, and Seth, Daniel L., Invariant Imbedding and Inverse Problems (1992)

Ramachandran, Vijaya, Bentley, Jon, Cole, Richard, Cunningham, William H., Guibas, Leo, King, Valerie, Lawler, Eugene, Lenstra, Arjen, Mulmuley, Ketan, Sleator, Daniel D., and Yannakakis, Mihalis, Proceedings of the Fourth Annual ACM-SIAM Symposium on Discrete Algorithms (1993)

Kleinman, Ralph, Angell, Thomas, Colton, David, Santosa, Fadil, and Stakgold, Ivar, Second International Conference on Mathematical and Numerical Aspects of Wave Propagation (1993)

Banks, H. T., Fabiano, R. H., and Ito, K., Identification and Control in Systems Governed by Partial Differential Equations (1993)

Sleator, Daniel D., Bern, Marshall W., Clarkson, Kenneth L., Cook, William J., Karlin, Anna, Klein, Philip N., Lagarias, Jeffrey C., Lawler, Eugene L., Maggs, Bruce, Milenkovic, Victor J., and Winkler, Peter, Proceedings of the Fifth Annual ACM-SIAM Symposium on Discrete Algorithms (1994)

Lewis, John G., Proceedings of the Fifth SIAM Conference on Applied Linear Algebra (1994)

Brown, J. David, Chu, Moody T., Ellison, Donald C., and Plemmons, Robert J., Proceedings of the Cornelius Lanczos International Centenary Conference (1994)

Dongarra, Jack J. and Tourancheau, B., Proceedings of the Second Workshop on Environments and Tools for Parallel Scientific Computing (1994)

Bailey, David H., Bjørstad, Petter E., Gilbert, John R., Mascagni, Michael V., Schreiber, Robert S., Simon, Horst D., Torczon, Virginia J., and Watson, Layne T., Proceedings of the Seventh SIAM Conference on Parallel Processing for Scientific Computing (1995)

Clarkson, Kenneth, Agarwal, Pankaj K., Atallah, Mikhail, Frieze, Alan, Goldberg, Andrew, Karloff, Howard, Manber, Udi, Munro, Ian, Raghavan, Prabhakar, Schmidt, Jeanette, and Young, Moti, Proceedings of the Sixth Annual ACM-SIAM Symposium on Discrete Algorithms (1995)

Cohen, Gary, Third International Conference on Mathematical and Numerical Aspects of Wave Propagation (1995)

Engl, Heinz W., and Rundell, W., GAMM–SIAM Proceedings on Inverse Problems in Diffusion Processes (1995)

Angell, T. S., Pamela Cook, L., Kleinman, R. E., and Olmstead, W. E., Nonlinear Problems in Applied Mathematics (1995)

Cook, Pamela L., Roytburd, Victor, and Tulin, Marshal, Mathematics Is for Solving Problems (1996)

Adams, Loyce and Nazareth, J. L., Linear and Nonlinear Conjugate Gradient-Related Methods (1996)

LINEAR AND NONLINEAR CONJUGATE GRADIENT-RELATED METHODS

Edited by **Loyce Adams**
University of Washington
Seattle, Washington

J. L. Nazareth
Washington State University
Pullman, Washington

Society for Industrial and
Applied Mathematics
Philadelphia

LINEAR AND NONLINEAR CONJUGATE GRADIENT-RELATED METHODS

Proceedings of the Conference on Linear and Nonlinear Conjugate Gradient-Related Methods, part of the Summer Research Conferences in the Mathematical Sciences. University of Washington, Seattle, Washington, July 9–13, 1995.

This conference was sponsored by the American Mathematical Society, the Institute of Mathematical Statistics, and the Society for Industrial and Applied Mathematics and was supported in part by a grant from the National Science Foundation.

Copyright © 1996 by the Society for Industrial and Applied Mathematics.

10 9 8 7 6 5 4 3 2 1

All rights reserved. Printed in the United States of America. No part of this book may be reproduced, stored, or transmitted without the written permission of the publisher. For information, write to the Society for Industrial and Applied Mathematics, 3600 University City Science Center, Philadelphia, PA 19104-2688.

Library of Congress Catalog Card Number: 96-68754

ISBN 0-89871-376-5

siam. is a registered trademark.

PREFACE

The purpose of the AMS–IMS–SIAM Summer Research Conference on *Linear and Nonlinear Conjugate Gradient-Related Methods* was to bring together leading researchers and practitioners in computational linear algebra and computational optimization, as well as younger postdocs and advanced graduate students, in order to foster communication and build bridges between linear and nonlinear conjugate gradient (CG)-related research (and its applications).

The term "conjugate gradient-related" in the title of the meeting is used generically, and with a broad interpretation, to connote an important class of iterative methods requiring limited computer storage. They are designed to solve large-scale (high-dimensional) problems involving a finite number of variables, in particular, systems of linear equations, linear least-squares problems, nonlinear optimization and nonlinear least-squares problems, and systems of nonlinear equations. Such problems are central to scientific computation. A particularly rich source arises from the discretization, in various ways, of infinite-dimensional problems, for example, systems of ordinary and partial differential equations, variational problems and optimal control problems.

Linear CG-related methods are techniques of computational linear algebra; nonlinear CG-related methods are techniques of computational optimization. The two areas share a broad common frontier, and one of the most easily traversed crossing points is via the following simple observation: solving a positive definite symmetric system of linear equations is equivalent to minimizing a strictly convex quadratic functional. This equivalence helped enrich early algorithmic advances in both subjects, facilitating, in particular, the original discovery of the CG method by Hestenes and Stiefel (and Lanczos) in 1952 and its adaptation to nonlinear optimization by Fletcher and Reeves in 1964. These developments gave rise to a very active research area of computational mathematics.

There has been an increasing divergence between researchers in CG-related methods for linear systems and CG-related methods for nonlinear optimization, and the two research communities are sometimes surprisingly isolated from one another, to the detriment of both. On the computational linear algebra side, we have CG/GMRES/Lanczos/..., with an emphasis on preconditioning techniques, nonsymmetric systems, and short recursions (to reduce cost and storage) that use lookahead strategies to avoid breakdown. Recently, there have also been attempts to explain how these methods behave in finite precision arithmetic. On the computational optimization side, we have nonlinear CG/Limited-Memory quasi-Newton (L-QN)/Truncated-Newton and Truncated QN/..., with an emphasis on convergence analysis, rate of convergence analysis, and implementation strategies. Both communities have a great deal to learn from one another. The need to strengthen synergistic connections between linear and nonlinear CG-related methods provided an organizing focus and theme for the conference.

The meeting took place on the campus of the University of Washington during a period of five days (Sunday through Thursday, July 9–13, 1995) with a half-day break on Tuesday afternoon. There were 10 one-hour talks by the principal invited speakers, and 20 half-hour talks by other invited participants. Altogether there were 53 registered attendees. Every attempt was made to alternate linear and nonlinear CG topics that we felt were related. This arrangement resulted in many discussions during the conference breaks between the two communities.

The invited presentations on the first day helped set the stage by introducing the subject matter from different perspectives: linear (O'Leary), nonlinear (Nocedal), differential geometry (Edelman/Smith), and applications (Young). These talks served as a good introduction for the students and young researchers who attended, but at the same time provided enough substantial and controversial material to ensure lively discussion throughout the meeting. The remaining talks in the conference represented a nice mixture of current research topics in both linear and nonlinear CG methods. Novel research approaches and results were presented and discussed.

Papers from the one-hour talks are included in the proceedings. In addition we gave the senior professors (emeritus professors) who participated an option to contribute shorter views-type articles.

The well-known templates book by Barrett et al. (published by SIAM in 1994), coupled with O'Leary's historical background article, gives comprehensive background information and references on the linear CG method and its variants. Because there is nothing analogous to the templates overview available for nonlinear CG methods, the article by Nazareth was included to give background and to supplement introductory information given in the article by Nocedal.

Collectively, the articles in this volume make many interesting observations and identify some of the important open research questions in the CG-related area. Thus, in particular,

- Dianne O'Leary observes that the themes that motivated the early researchers in the 1950s are still the unanswered topics today, namely, the near-breakdown and loss of orthogonality in nonsymmetric conjugate gradient iterations, the behavior of CG on ill-posed problems, the selection of a good preconditioner, and the choice of the form of the recursions.

- Jorge Nocedal describes three key topics at the frontier of nonlinear CG research, including a new approach (Discrete Newton Method with Memory) that fuses the properties of CG and quasi-Newton methods.

- Alan Edelman and Steve Smith observe that differential geometry can provide a different viewpoint for solving eigen-like problems.

- David P. Young and his coworkers argue that black-box nonlinear optimization codes are too restrictive in some application work and demonstrate how nonlinear optimization can be successfully applied in a setting that arises at Boeing.

- Andy Conn and his collaborators describe a new approach to large-scale optimization that relies on minimizations in low-dimensional iterated subspaces. The latter are defined by conjugate directions generated via an (inner) preconditioned linear CG algorithm.

- Bill Davidon gives a succinct account of conjugacy relations for conic generalizations of quadratic functions. There seem to be unexplored overlaps between this type of approach and ideas from differential geometry described by Edelman/Smith.

- Anne Greenbaum shows that it remains an open question to provide *a priori* bounds for the convergence of the GMRES method in terms of some simple characteristic properties of the coefficient matrix. She also states that nonoptimal Krylov space methods that are based on short recursions have not been related to the optimal GMRES method. In fact, she states that "perhaps the greatest remaining open problem in this field is to find a method that generates provable 'near-optimal' approximations in some standard norm, while still requiring $O(n)$ work and storage (in addition to the matrix-vector multiplication) at each iteration—or to prove that such a method does not exist."

- Michael Saunders gives a new regularization technique for solving augmented systems associated with least-squares and interior (barrier) methods of linear programming, which has important implications for stability.

- Laurence Dixon reconsiders the properties of the CG method and gives new results on retaining prior information, for example, after a restart.

- Teri Barth and Thomas Manteuffel examine the question of the existence of a CG method based on short multiple recursions that is optimal for nonsymmetric problems. They give sufficent conditions on the class of matrices for which such a method exists.

- Sanjay Mehrotra and Jen-Shan Wang show how to effectively precondition the linear CG algorithm when it is used to compute search directions within an interior-point method for network flow problems. The queston of how to effectively employ iterative methods, and in particular the CG method, within interior-point techniques for the *general* linear programming problem is still very much an open issue.

- Michael DeLong and James Ortega discuss the issues of finding a good preconditioner for nonsymmetric problems. They show that the SOR method can be effective on parallel computers for these problems.

- Finally, Larry Nazareth gives a broad overview of past developments and future trends in nonlinear CG-related research.

We want to acknowledge the support provided by the National Science Foundation that made the meeting possible. We would also like to thank the American Mathematical Society and, in particular, Wayne Drady for the wonderful support that was provided both before and during the meeting. Susan Ciambrano and Bernadetta DiLisi at SIAM have been extremely helpful in making the proceedings for this meeting become a reality. Finally we thank all the speakers and other participants for helping to make the meeting a success.

Loyce Adams:
adams@amath.washington.edu

Larry Nazareth:
nazareth@amath.washington.edu

COMPLETE LIST OF SPEAKERS, E-MAIL ADDRESSES, AND TITLES OF PRESENTATIONS

Barth, Terri
tbarth@carbon.cudenver.edu
Variable Metric Conjugate Gradient Methods

Burkhart, Richard
burkhart@atc.boeing.com
Inexact Newton Methods in Real Time Signal Processing

Chan, Tony
chan@math.ucla.edu
Iterative Methods for Total Variation Image Reconstruction

Conn, Andy
arconn@watson.ibm.com
Aspects of CG-Related Methods in Nonlinear Optimization

Davidon, Bill
wdavidon@haverford.edu
Conjugate Directions in Perspective

DeLong, Michael/J. Ortega
mike@virginia.edu
SOR as a Preconditioner

Dixon, Laurence
L.C.Dixon@herts.ac.uk
Memory Retention by Conjugate Gradient Methods

Edelman, Alan
edelman@math.mit.edu
Conjugate Gradient on the Stiefel and Grassman Manifolds: Numerical Linear Algebra meets Differential Geometry (Part I)

Greenbaum, Anne
greenbau@greenbaum.cims.nyu.edu
Recent Advances and Open Problems in Iterative Methods for Solving Linear Systems

Leonard, Mike/P. Gill
mwl@sdna3.ucsd.edu
Large Scale Quasi-Newton Methods using Limited Storage

Manteuffel, Tom
tmanteuf@newton.colorado.edu
Conjugate Gradients Using Multiple Recursion

Mehrotra, Sanjay
mehrotra@iems.nwu.edu
Conjugate Gradient Based Implementation of Interior-Point Methods for Network Flow Problems

Nachtigal, Noel
santa@cs.utk.edu
QMRPACK and Applications

Nash, Stephen
snash@gmuvax.gmu.edu
Preconditioning Reduced Matrices

Nocedal, Jorge
nocedal@eecs.nwu.edu
CG Methods in Nonlinear Optimization

Nikishin, Andy/Alex Yeremin
n3661@cray.com
Variable Block CG Methods

O'Leary, Dianne
oleary@mimsy.umd.edu
Conjugate Gradients: Images, Parallelism, and Preconditioning

Parlett, Beresford
parlett@math.berkeley.edu
Approximate Solutions from Krylov Subspaces: Representations and Harmonic Ritz Values

Pilla, Ramani/B.G. Lindsay
ramani@stat.psu.edu
Conjugate Gradient Method: Application to Neural Networks

Saunders, Michael
mike@sol-michael.stanford.edu
CG and LDL' Methods for Sparse Least Squares

Saylor, Paul
saylor@sal.cs.uiuc.edu
Estimating the Error in CG

Smith, Steve
stsmith@ll.mit.edu
Conjugate Gradient on the Stiefel and Grassman Manifolds: Numerical Linear Algebra meets Differential Geometry (Part II)

Starke, Gerhard
starke@newton.colorado.edu
Krylov Subspace Iterations in Multilevel Methods for Nonsymmetric Problems

Thelen, Brian
thelen@erim.org
Large-scale Optimization Problems related to Techniques for Simultaneous Aberration Estimation and Image Deblurring

Tseng, Paul
tseng@math.washington.edu
Non-monotone Gradient Methods for Minimizing the Sum of Differentiable Functions, with Applications to Training Nonlinear Neural Networks

Tuminaro, Ray
tuminaro@gaston.cs.sandia.gov
AZTEC: An EASY-TO-USE Parallel Iterative Solver Package for the Solution of Linear Systems Arising from Unstructured Problems

Watkins, David
watkins@amath.washington.edu
Short Recursions Associated with Unitary Operators

Ye, Qiang/Charles Tong
matong@uxmail.ust.hk
Finite Precision Analysis of the Biconjugate Gradient Method

Young, David M.
young@cs.utexas.edu
Orthogonal-Based Iterative Methods for Nonsymmetric Linear Systems

Young, David P.
dpy6629@cfdd51.ca.boeing.com
Requirements for Optimization in Engineering

CONTENTS

1   Conjugate Gradients and Related KMP Algorithms: The Beginnings
Dianne P. O'Leary

9   Conjugate Gradient Methods and Nonlinear Optimization
Jorge Nocedal

24   On Conjugate Gradient-Like Methods for Eigen-Like Problems
Alan Edelman and Steven T. Smith

37   Requirements for Optimization in Engineering
D.P. Young, W.P. Huffman, R.G. Melvin, F.T. Johnson, C.L. Hilmes, and M.B. Bieterman

50   On Iterated-Subspace Minimization Methods for Nonlinear Optimization
A.R. Conn, Nick Gould, A. Sartenaer, and Ph. L. Toint

79   Conjugate Directions in Perspective
William C. Davidon

83   Krylov Subspace Approximations to the Solution of a Linear System
Anne Greenbaum

92   Cholesky-based Methods for Sparse Least Squares: The Benefits of Regularization
Michael A. Saunders

101   Memory Retention in Conjugate Gradient Methods
Laurence Dixon

107   Conjugate Gradient Algorithms Using Multiple Recursions
Teri L. Barth and Thomas A. Manteuffel

124   Conjugate Gradient Based Implementation of Interior Point Methods for Network Flow Problems
Sanjay Mehrotra and Jen-Shan Wang

143 SOR as a Parallel Preconditioner
M.A. DeLong and J.M. Ortega

149 A View of Conjugate Gradient-Related Algorithms for Nonlinear Optimization
J.L. Nazareth

Chapter 1
Conjugate Gradients and Related KMP Algorithms: The Beginnings[*]

Dianne P. O'Leary[†]

Abstract

In the late 1940's and early 1950's, newly available computing machines generated intense interest in solving "large" systems of linear equations. Among the algorithms developed were several related methods, all of which generated bases for Krylov subspaces and used the bases to minimize or orthogonally project a measure of error. These methods include the conjugate gradient algorithm and the Lanczos algorithm. We refer to these algorithms as *the KMP family* and discuss its origins, emphasizing research themes that continue to have central importance.

1 Introduction

The conjugate gradient algorithm is now close to 45 years old. As this algorithm and related ones have grown in importance and widespread use, some mythology has developed regarding the early history of the family. The purpose of this overview is to discuss the origins of this family of algorithms, emphasizing research themes that continue to have central importance in current research.

One difficulty in talking about these algorithms is in terminology. Our discussion concerns certain methods that generate a basis for a *Krylov subspace*, the subspace spanned by the vectors b, Ab, ..., $A^{k-1}b$, where A is a given matrix and b is a given vector. "Conjugate gradient family" is too restrictive, since it implies a notion of conjugacy that is lost in some of the nonsymmetric algorithms. "Krylov subspace family" is too broad, since it rightly includes stationary iterative methods such as SOR, and nonstationary methods such as the Chebyshev semi-iterative method. The "Lanczos family" appears to exclude Arnoldi-based algorithms. One characterization of these methods is that they use the Krylov subspace either to minimize an error function (e.g., conjugate gradients), to project the original operator (e.g., the Lanczos algorithm for computing eigenvalues), or to project the error in the solution into a progressively smaller-dimensional subspace (e.g., the Arnoldi method). For lack of a better term, we will use the phrase *KMP* family to denote those Krylov-subspace generating algorithms that either *M*inimize an error or *P*roject an operator or error. This includes the methods of Lanczos, Arnoldi, conjugate gradients, GMRES, CGS, QMR, and an alphabet soup of other algorithms.

The following sections discuss some of the key early steps in the development of the KMP family. A reader interested in recent developments could begin with [24], [3], and the papers in this volume.

---

[*]This work was supported by the National Science Foundation under Grant CCR-95-03126.
[†]Department of Computer Science and Institute for Advanced Computer Studies, University of Maryland, College Park, MD 20742 (oleary@cs.umd.edu).

2 The Beginnings

Researchers in the late 1940's and early 1950's were concerned with developing effective algorithms for solving basic linear algebra problems (systems of linear equations and eigenvalue computations) on the computing machines available after the end of World War II. These machines represented a revolutionary step forward, but had severe limits in power. Perhaps the most significant limitation was the amount of memory. A branch of the United States National Bureau of Standards (NBS), the Institute for Numerical Analysis (INA) within the National Applied Mathematics Laboratories, had been established in Los Angeles to study means by which important computational problems could be solved on such machines. The list of people associated with the INA includes Olga Taussky, John Todd, and others mentioned later in this section.

2.1 KMP Methods: Direct vs. Iterative

One myth concerning the origins of the conjugate gradient algorithm is that it was developed as a direct method rather than an iterative one. History does not provide much support for this view, however.

The study of iterative methods for eigenvalue problems and for linear systems was a central research theme at the INA, and a landmark paper was published by Cornelius Lanczos in 1950 in the *Journal of Research of the National Bureau of Standards* [36]. In this work, Lanczos developed a three-term recurrence relation for matrix polynomials (equivalently, for a basis for a Krylov subspace) and a biorthogonalization method for finding eigenvalues of nonsymmetric matrices.

Lanczos' idea was further developed by W. E. Arnoldi, who reinterpreted the algorithm and then derived a new one, reducing a general matrix to upper Hessenberg form [1]. Arnoldi presented the algorithm as iterative in nature, and discussed filtering the right-hand side before generating the Krylov sequence.

The earliest report of NBS activity on the conjugate gradient algorithm for solving systems of equations involving symmetric positive definite matrices was in an abstract for the Summer Meeting of the American Mathematical Society held in Minneapolis in September, 1951 [21]. This discussion of the three-term recurrence form of conjugate gradients is coauthored by George Forsythe, Magnus Hestenes, and J. Barkley Rosser. Hestenes wrote a technical report on this work in July, 1951 [31]. In it, he acknowledged discussions with "Forsythe, Lanczos, Paige, Rosser, Stein, and others." He specifically credited Forsythe and Rosser for the 3-term recurrence and L. J. Paige for the usual 2-term recurrence. The conjugate direction algorithm is somewhat older, discussed in a 1948 paper by Leslie Fox, H. D. Huskey, and Jim Wilkinson [22].

Meanwhile, Eduard Stiefel of E.T.H., Zurich, visited the INA, and presented a paper on the conjugate gradient algorithm at a workshop in August, 1951. His description of the "n-step iteration" was published in 1952 and noted the connection with the Lanczos (1950) work [49].

Hestenes and Stiefel decided to combine their efforts, and they published *the* conjugate gradient paper in 1952. They described conjugate gradients as "an iterative method that terminates" [32, p.410]. They devoted several pages to showing the relation between conjugate gradients and Gaussian elimination, but the bulk of the paper focused on the iterative properties, such as monotonicity of various error measures, methods for smoothing the initial residual, remedies for loss of orthogonality, the algebraic framework for preconditioning, and the relation to the Lanczos (1950) algorithm and continued fractions.

They also discussed determining the solution when the matrix A is rank deficient. They reported the use of the algorithm to solve 106 "difference equations" in 90 iterations. By 1958, researchers had solved Laplace's equation over a 10×10 grid in 11 Chebyshev iterations plus 2 conjugate gradient iterations [17].

Lanczos also published a paper in 1952 concerning iterative methods for linear systems [37]. He developed the use of his 1950 algorithm, and developed 2-term recurrences of biorthogonal polynomials.

The conjugate gradient iteration was extended to Hilbert space in the Ph.D. dissertation of R. M. Hayes, a U.C.L.A. student working at the INA [29]. Hayes established a linear convergence rate for general operators, with superlinear convergence if the operator is of the form identity plus a completely continuous operator. The roots of the idea of clustering eigenvalues to improve the convergence of conjugate gradients seem to date from here.

Clearly early researchers of the KMP family recognized the algorithms' usefulness as iterative methods and devoted considerable thought to *preconditioning* – either by rescaling the matrix or by filtering the initial error by the use of other iterative methods as Lanczos suggested in 1952 [37, p.45].

2.2 Assigning Credit

Considerable controversy has arisen regarding proper credit for the conjugate gradient algorithm. Lanczos's papers clearly focus on KMP algorithms for nonsymmetric matrices, while Hestenes and Stiefel limit their attention to Hermitian positive definite matrices. This extra restriction enabled the use of many additional important properties.

The claims of Lanczos, Hestenes, and Stiefel do not really overlap. Lanczos says in 1952 [37, p.53], "The latest publication of Hestenes [1951] and of Stiefel [1952] is closely related to the p, q algorithm of the present paper, although developed independently and from different considerations." Stiefel's claim in 1952 [49, p.23] was that, "After writing up the present work, I discovered on a visit to the Institute for Numerical Analysis (University of California) that these results were also developed somewhat later by a group there." Hestenes and Stiefel give a more complete citation in the 1952 paper [32, pp.409-410]:

> The method of conjugate gradients was developed independently by E. Stiefel of the Institute of Applied Mathematics at Zurich and by M. R. Hestenes with the cooperation of J. B. Rosser, G. Forsythe, and L. Paige of the Institute for Numerical Analysis, National Bureau of Standards. The present account was prepared jointly by M. R. Hestenes and E. Stiefel during the latter's stay at the National Bureau of Standards. The first papers on this method were given by E. Stiefel [1952] and M. R. Hestenes [1951]. Reports on this method were given by E. Stiefel and J. B. Rosser at a Symposium on August 23-25, 1951. Recently, C. Lanczos [1952] developed a closely related routine based on his earlier paper on eigenvalue problem. Examples and numerical tests of the method have been by R. Hayes, U. Hochstrasser, and M. Stein.

All of this supports Hestenes' memories in 1987 [30]: "I believe it was done in the following order. 1. Stiefel because he had carried out some large experiments which surely took place more than a month before he came to UCLA. I invented it within a month of his arrival. 2. Hestenes. 3. Lanczos. He is third because he would have been talking about it prior to my invention of the routine. I am sure that when he saw my paper he said to himself, 'I knew it all along'. The remarkable thing is that it took two years of study of iterative methods at INA before the cg-algorithm was devised."

It is clear that each of these people came upon the conjugate gradient algorithm from different considerations: Hestenes, from his long-time interest in conjugacy, dating back to a joint paper with G. D. Birkhoff in the 1930's; Stiefel from "simultaneous relaxation," adding a linear combination of several vectors to the current iterate; and Lanczos from polynomial recurrences. This multitude of roots for the KMP family is key to its richness of theoretical properties and its computational usefulness.

3 The 1960's

Another myth about the KMP family is that the algorithms were forgotten in the 1960's, discarded because they could not compete with Gauss elimination as a direct method. Although the algorithms were not much in favor among numerical mathematicians, they did achieve considerable success in applications in spectral analysis [6], lens design [19], geodesy [16], polar circulation [7], infrared spectral analysis [45], optimal control [48], collision theory [18], structural analysis [23], pattern recognition [39], power system load flow [53], optimal flight paths [5], nonrelativistic scattering [25], network analysis [35], and nuclear shell computation [47].

Some important work was also occurring in extending the usefulness of the KMP family. Wachspress showed the effectiveness of the algorithm preconditioned by alternating direction implicit methods (ADI) in solving discretizations of partial differential equations [52]. Golub and Kahan discussed the use of the Lanczos algorithm in computing the singular value decomposition [26]. Fletcher and Reeves extended the conjugate gradient algorithm to minimization of non-quadratic functions [20], opening an important area of research. Somewhat earlier, Davidon [14] had developed the first algorithm in the quasi-Newton family, but it had not yet been recognized as a relative of the conjugate gradient algorithm that generates the same sequence of iterates when applied to quadratic minimization.

Major steps in understanding the convergence behavior of the conjugate gradient algorithm were made by Kaniel [34] and Daniel [12]. Although both papers contained some errors [4, 13, 8], the standard bounds on the convergence rate for conjugate gradients, derived using Chebyshev polynomials, can be found here.

Thus the 1960's brought considerable progress in the use and understanding of the KMP family.

4 The 1970's

The key paper reviving the interest of numerical analysts in the investigation of KMP methods was a 1970 presentation, at an Oxford conference sponsored by the Institute of Mathematics and Applications, by John Reid of the U.K. Atomic Energy Research Establishment, Harwell [46]. Reid compared the numerical performance of several variants of the conjugate gradient algorithm and emphasized the algorithm's usefulness on well-conditioned problems.

This stimulated interest among a number of researchers, and activity in the area blossomed. It was clear that progress was needed on several fronts: preconditioners that would turn ill-conditioned problems into well-conditioned ones, stabilized forms of the algorithms, and extensions to broader problem classes.

The revival of interest in preconditioning produced the important paper of J. Meijerink and Henk van der Vorst [38], available in preprint form in the early 1970's. They developed an algorithm for computing an incomplete LU factorization of an M-matrix (as did Richard Varga in a 1960 paper [51]). This work inspired the hope of having a library of preconditioners that would apply to broad problem classes. Preconditioning was also discussed by Owe Axelsson [2], by Paul Concus, Gene Golub, and Dianne O'Leary [10], and for nonlinear problems by Jim Douglas and Todd Dupont [15].

Two important extensions of the conjugate gradient algorithm were given during this period. Paul Concus, Gene Golub, and Olof Widlund solved problems in which the Hermitian part of the matrix was positive definite and could be used as a preconditioner [9, 54]. Chris Paige and Michael Saunders showed how to compute iterates in case a matrix was indefinite [44], resulting in SYMMLQ and related algorithms.

Research on the eigenvalue problem also contributed to making the KMP algorithms more useful for solving linear systems. Chris Paige [43, 40, 41, 42], Beresford Parlett and W. Kahan [33], and Jane Cullum and Ralph Willoughby [11] all made important contributions to understanding the behavior of the Lanczos recursion under inexact arithmetic. Gene Golub, Richard Underwood, and Jim Wilkinson developed a block form of the Lanczos algorithm [28, 50] that later inspired the development of block KMP algorithms.

5 Closing Comments

Further information on the early history of the KMP family, as well as a much more complete bibliography, can be found in [27].

Perhaps the most important lesson to be learned in scanning the early literature on KMP methods is that the themes that motivated the earliest work continue to this day to be basic questions that have not been fully answered.

- The influence of inexact arithmetic led the researchers of the 1950's to develop and test various forms of the iterations, and researchers of the 1990's still discuss how to handle near-breakdown of the nonsymmetric iterations and the influence of loss of othogonality of the basis vectors.

- Important aspects of the convergence behavior of the conjugate gradient iteration were understood relatively early; important questions still remain open concerning its behavior on discretizations of ill-posed problems, even though Lanczos himself had a partial understanding [37]. Convergence behavior of the nonsymmetric iterations is not nearly as well understood.

- Preconditioning remains the key to effective use of the KMP family of algorithms, and the search for preconditioners effective on broad problem classes continues.

Future work in all of these areas will continue to be inspired by the fundamental questions raised by the earliest researchers.

6 Acknowledgements

I am grateful to Loyce Adams and Larry Nazareth for their excellent work in organizing the conjugate gradient workshop and for inviting this survey.

References

[1] W. E. Arnoldi, *The principle of minimized iterations in the solution of the matrix eigenvalue problem*, Quarterly of Appl. Math., 9 (1951), pp. 17–29.

[2] O. Axelsson, *A generalized SSOR method*, BIT, 12 (1972), pp. 443–467.

[3] R. Barrett, M. Berry, T. F. Chan, J. Demmel, J. Donato, J. Dongarra, V. Eijkhout, R. Pozo, C. Romine, and H. van der Vorst, *Templates for the Solution of Linear Systems*, SIAM, Philadelphia, 1993.

[4] G. G. Belford and E. H. Kaufman,Jr., *An application of approximation theory to an error estimate in linear algebra*, Math. of Comp., 28 (1974), pp. 711–712.

[5] B. L. Bierson, *A discrete-variable approximation to optimal flight paths*, Astronautica Acta, 14 (1969), pp. 157–169.

[6] A. A. Bothner-By and C. Naar-Colin, *The proton magnetic resonance spectra of 2,3-disubstituted n-butanes*, J. of the ACS, 84 (1962), pp. 743–747.

[7] W. J. Campbell, *The wind-driven circulation of ice and water in a polar ocean*, J. of Geophysical Research, 70 (1965), pp. 3279–3301.

[8] A. I. Cohen, *Rate of convergence of several conjugate gradient algorithms*, SIAM J. Numer. Anal., 9 (1972), pp. 248–259.

[9] P. Concus and G. H. Golub, *A generalized conjugate gradient method for nonsymmetric systems of linear equations*, in Computing Methods in Applied Sciences and Engineering (2nd Internat. Symposium, Versailles 1975), Part I, R. Glowinski and J. L. Lions, eds., Springer Lecture Notes in Econ. and Math. Systems 134, New York, 1976, pp. 56–65.

[10] P. Concus, G. H. Golub, and D. P. O'Leary, *A generalized conjugate gradient method for the numerical solution of elliptic partial differential equations*, in Sparse Matrix Computations, J. R. Bunch and D. J. Rose, eds., Academic Press, New York, 1976, pp. 309–332.

[11] J. Cullum and R. A. Willoughby, *Lanczos and the computation in specified intervals of the spectrum of large, sparse real symmetric matrices*, in Sparse Matrix Proceedings 1978, I. S. Duff and G. W. Stewart, eds., SIAM, Philadelphia, 1979, pp. 220–255.

[12] J. W. Daniel, *The conjugate gradient method for linear and nonlinear operator equations*, SIAM J. Numer. Anal., 4 (1967a), pp. 10–26.

[13] J. W. Daniel, *A correction concerning the convergence rate for the conjugate gradient method*, SIAM J. Numer. Anal., 7 (1970b), pp. 277–280.

[14] W. C. Davidon, *Variable metric method for minimization*, tech. rep., ANL-5990, Argonne National Laboratory, Argonne, Illinois, 1959.

[15] J. Douglas, Jr. and T. Dupont, *Preconditioned conjugate gradient iteration applied to Galerkin methods for a mildly-nonlinear Dirichlet problem*, in Sparse Matrix Computations, J. R. Bunch and D. J. Rose, eds., Academic Press, New York, 1976, pp. 333–348.

[16] H. M. Dufour, *Resolution des systemes lineaires par la methode des residus conjugues*, Bulletin Géodésique, 71 (1964), pp. 65–87.

[17] M. Engeli, T. Ginsburg, H. Rutishauser, and E. Stiefel, *Refined Iterative Methods for Computation of the Solution and the Eigenvalues of Self-Adjoint Boundary Value Problems*, Birkhauser Verlag, Basel/Stuttgart, 1959.

[18] B. E. Eu, *Method of moments in collision theory*, J. Chem. Phys., 48 (1968), pp. 5611–5622.

[19] D. P. Feder, *Automatic lens design with a high-speed computer*, J. of the Optical Soc. of Amer., 52 (1962), pp. 177–183.

[20] R. Fletcher and C. M. Reeves, *Function minimization by conjugate gradients*, Computer J., 7 (1964), pp. 149–154.

[21] G. E. Forsythe, M. R. Hestenes, and J. B. Rosser, *Iterative methods for solving linear equations*, Bull. Amer. Math. Soc., 57 (1951), p. 480.

[22] L. Fox, H. D. Huskey, and J. H. Wilkinson, *Notes on the solution of algebraic linear simultaneous equations*, Quart. J. of Mech. and Appl. Math., 1 (1948), pp. 149–173.

[23] R. L. Fox and E. L. Stanton, *Developments in structural analysis by direct energy minimization*, AIAA J., 6 (1968), pp. 1036–1042.

[24] R. W. Freund, G. H. Golub, and N. M. Nachtigal, *Iterative solution of linear systems*, Acta Numerica, (1991), pp. 57–100.

[25] C. R. Garibotti and M. Villani, *Continuation in the coupling constant for the total K and T matrices*, Il Nuovo Cimento, 59 (1969), pp. 107–123.

[26] G. Golub and W. Kahan, *Calculating the singular values and pseudo-inverse of a matrix*, SIAM J. Numer. Anal., 2(Series B) (1965), pp. 205–224.

[27] G. H. Golub and D. P. O'Leary, *Some history of the conjugate gradient and Lanczos algorithms: 1948-1976*, SIAM Review, 31 (1989), pp. 50–102.

[28] G. H. Golub, R. Underwood, and J. H. Wilkinson, *The Lanczos algorithm for the symmetric*

$Ax = \lambda Bx$ *problem*, tech. rep., Stanford University Computer Science Department, Stanford, California, 1972.
[29] R. M. Hayes, *Iterative methods of solving linear problems on Hilbert space*, in Contributions to the Solution of Systems of Linear Equations and the Determination of Eigenvalues, O. Taussky, ed., vol. 39 Applied Mathematics Series, National Bureau of Standards, U.S. Government Printing Office, Washington, D.C., 1954, pp. 71–103.
[30] M. Hestenes, *Private communcation*, (July 1, 1987).
[31] M. R. Hestenes, *Iterative methods for solving linear equations*, tech. rep., NAML Report 52-9, July 2, 1951, National Bureau of Standards, Los Angeles, California, 1951. Reprinted in *J. of Optimization Theory and Applications* 11 (1973), pp. 323-334.
[32] M. R. Hestenes and E. Stiefel, *Methods of conjugate gradients for solving linear systems*, J. Res. Nat. Bur. Standards, 49 (1952), pp. 409–436.
[33] W. Kahan and B. N. Parlett, *How far should you go with the Lanczos process?*, in Sparse Matrix Computations, J. R. Bunch and D. J. Rose, eds., Academic Press, New York, 1976, pp. 131–144.
[34] S. Kaniel, *Estimates for some computational techniques in linear algebra*, Math. of Comp., 95 (1966), pp. 369–378.
[35] K. Kawamura and R. A. Volz, *On the convergence of the conjugate gradient method in Hilbert space*, IEEE Trans. on Auto. Control, AC-14 (1969), pp. 296–297.
[36] C. Lanczos, *An iteration method for the solution of the eigenvalue problem of linear differential and integral operators*, J. Res. Nat. Bur. Standards, 45 (1950), pp. 255–282.
[37] C. Lanczos, *Solution of systems of linear equations by minimized iterations*, J. Res. Nat. Bur. Standards, 49 (1952), pp. 33–53.
[38] J. Meijerink and H. A. V. der Vorst, *An iterative solution method for linear systems of which the coefficient matrix is a symmetric M-matrix*, Math. Comp., 31 (1977), pp. 148–162.
[39] G. Nagy, *Classification algorithms in pattern recognition*, IEEE Trans. on Audio and Electroacoustics, AU-16 (1968a), pp. 203–212.
[40] C. C. Paige, *Practical use of the symmetric Lanczos process with re-orthogonalization*, BIT, 10 (1970), pp. 183–195.
[41] ———, *The computation of eigenvalues and eigenvectors of very large sparse matrices*, tech. rep., Ph. D. dissertation, University of London, 1971.
[42] ———, *Computational variants of the Lanczos method for the eigenproblem*, J. Inst. Maths. Applics., 10 (1972), pp. 373–381.
[43] ———, *Error analysis of the Lanczos algorithm for tridiagonalizing a symmetric matrix*, J. Inst. Maths. Applics., 18 (1976), pp. 341–349.
[44] C. C. Paige and M. A. Saunders, *Solution of sparse indefinite systems of linear equations*, SIAM J. Numer. Anal., 12 (1975), pp. 617–629.
[45] J. Pitha and R. N. Jones, *A comparison of optimization methods for fitting curves to infrared band envelopes*, Canadian J. of Chemistry, 44 (1966), pp. 3031–3050.
[46] J. K. Reid, *On the method of conjugate gradients for the solution of large sparse systems of linear equations*, in Large Sparse Sets of Linear Equations, Academic Press, New York, 1971, pp. 231–254.
[47] T. Sebe and J. Nachamkin, *Variational buildup of nuclear shell model bases*, Annals of Physics, 51 (1969), pp. 100–123.
[48] J. F. Sinnott, Jr. and D. G. Luenberger, *Solution of optimal control problems by the method of conjugate gradients*, in 1967 Joint Automatic Control Conference, Preprints of Papers, Lewis Winner, New York, 1967, pp. 566–574.
[49] E. Stiefel, *Über einige methoden der relaxationsrechnung*, Zeitschrift für angewandte Mathematik und Physik, 3 (1952), pp. 1–33.
[50] R. Underwood, *An iterative block Lanczos method for the solution of large sparse symmetric eigenproblems*, tech. rep., Ph.D. dissertation, Stanford University Computer Science Dept. Report STAN-CS-75-496, Stanford, California, 1975.
[51] R. S. Varga, *Factorization and normalized iterative methods*, in Boundary Problems in Differential Equations, R. E. Langer, ed., University of Wisconsin Press, Madison, 1960,

pp. 121–142.
[52] E. L. Wachspress, *Extended application of alternating direction implicit iteration model problem theory*, J. Soc. Industr. Appl. Math., 11 (1963), pp. 994–1016.
[53] Y. Wallach, *Gradient methods for load-flow problems*, IEEE Trans. on Power Apparatus and Systems, PAS-87 (1968), pp. 1314–1318.
[54] O. Widlund, *A Lanczos method for a class of nonsymmetric systems of linear equations*, SIAM J. Numer. Anal., 15 (1978), pp. 801–812.

Chapter 2
Conjugate Gradient Methods and Nonlinear Optimization

Jorge Nocedal*

Abstract

This paper begins with a brief history of the conjugate gradient method in nonlinear optimization. A challenging problem arising in meteorology is then presented to illustrate the kinds of large-scale problems that need to be solved at present. The paper then discusses three current areas of research: the development and analysis of nonlinear CG methods, the use of the linear conjugate gradient method as an iterative linear solver in Newton-type methods, and the design of new algorithms for large-scale optimization that make use of the interplay between quasi-Newton and conjugate gradient methods.

Key words: conjugate gradient method, quasi-Newton method, nonlinear optimization, large-scale optimization, constrained optimization, nonlinear conjugate gradient method

1 Introduction.

The conjugate gradient (CG) method has always played a special role in nonlinear optimization. It is related to quasi-Newton methods in many interesting ways that are being investigated to this day with the purpose of designing faster algorithms for large-scale optimization. The conjugate gradient method can also be modified to produce a class of algorithms called nonlinear conjugate gradient methods that possess unique properties among optimization methods. In addition, the conjugate gradient method can be used as an iterative linear solver in implementations of Newton and quasi-Newton methods, and by fully exploiting the subspace minimization properties of CG, these implementations give rise to robust, economical and rapidly convergent optimization methods. In this paper we will discuss a few of the areas in which CG plays an important role in optimization.

But let us begin with a brief historical account of the conjugate gradient method in nonlinear optimization. The story begins with Davidon's invention of quasi-Newton (or variable metric) methods in the late 1950s. Unknowing of the existence of the conjugate gradient method, Davidon [6] proposed an algorithm for nonlinear optimization that possesses a fast rate of convergence and finite termination on quadratic objective functions. A few years later Fletcher and Powell [15] showed that this algorithm is equivalent to the conjugate gradient method when applied, with exact line searches, to convex quadratic functions; the algorithm thus came to be known as the Davidon-Fletcher-Powell (DFP)

---

*Department of Electrical Engineering and Computer Science, Northwestern University, Evanston Il 60208-3118. E-mail: nocedal@eecs.nwu.edu. This author was supported by National Science Foundation Grants CCR-940081 and ASC-9213149, and by Department of Energy Grant DE-FG02-87ER25047-A004.

method. During the next ten years, several refinements and variations of the DFP method gave rise to the very effective quasi-Newton algorithms used today with great success in a great variety of areas of application. The very popular BFGS method is a direct descendent of the DFP method.

Almost immediately after the publication of the DFP method, Fletcher and Reeves [16] proposed another algorithm for nonlinear optimization that appeared to be even more closely related to the conjugate gradient method. Unlike quasi-Newton methods, the algorithm of Fletcher and Reeves does not require matrix storage and is very similar in form to the conjugate gradient method. It was the first nonlinear conjugate gradient method, and subsequent research showed that a simple variation due to Polak and Ribière gives good practical performance. Nonlinear conjugate gradient methods are designed so as to be equivalent to the (linear) conjugate gradient method when applied to quadratic functions.

The DFP and Fletcher-Reeves methods marked the beginning of a new era in nonlinear optimization, and much of the research performed during the last 30 years is directly related to these two seminal algorithms. It has been shown [34], [29], [29],[3], [31], [21] that quasi-Newton and nonlinear conjugate gradient methods can be related in various ways, most notably by introducing adaptive preconditioning techniques in the nonlinear CG methods. Research performed during the 1980s showed that there is a class of algorithms that fills the gap between quasi-Newton and conjugate gradient methods, and considerable effort was devoted to finding an algorithm with the right balance between these two approaches. It turned out that the two most successful algorithms for large-scale optimization that emerged in the 1980s – quasi-Newton methods for partially separable optimization and limited memory methods – lie completely in the domain of quasi-Newton methods. Thus the pendulum has swung towards the quasi-Newton approach, and at present, nonlinear conjugate gradient methods do not play a dominant role in numerical optimization. We will return to this in §3.

Nevertheless the *linear* conjugate gradient method continues to gain importance in the implementation of Newton-type methods for both constrained and unconstrained optimization, as will be discussed in §4. In addition, the interplay between the linear conjugate gradient method and nonlinear optimization algorithms is currently being explored with the goal of designing more robust and cost-effective algorithms for large-scale optimization; we will discuss this in §5.

This paper is not a survey of recent work in conjugate gradient methods for optimization. It focuses only on three important areas of optimization where the conjugate gradient method is currently playing a central role, and that are indicative of future research directions. To motivate the discussion that follows, and to set the stage for the new algorithms discussed in §5, we first describe an important practical problem that illustrates the tight restrictions under which large-scale optimization algorithms must perform.

2   An Example: Assimilation of Data in Meteorology.

There is a demand for solving increasingly larger nonlinear optimization problems under strict limitations of storage and computing time. An example of this arises in the four-dimensional variational assimilation of data in meteorology – an inverse problem that we now describe in some detail.

A major obstacle in weather forecasting is our inability to determine the initial state of the atmosphere with sufficient detail to resolve all the degrees of freedom of physical interest. Even though much more information is now available due to remotely sensed observations from satellites, it is far from sufficient, and it is therefore necessary to develop numerical methods that extract the maximum amount of information from the data.

Four-dimensional variational assimilation is a novel technique for doing this. It is a least squares technique in which the objective is to minimize, with respect to the set of initial conditions, the distance between the fields predicted by the model and the fields measured experimentally, while constraining the variables to exactly satisfy the equations of motion [7], [22], [24]. This represents a large-scale nonlinear optimization problem whose objective function we now describe.

Suppose that we have a set of observation vectors $\mathbf{Z_i}$ taken at times τ_i, $1 \leq i \leq n$. Each observation vector consists of a set of meteorological data, such as humidity, wind strength and direction, vorticity and temperature. Suppose also that we have a numerical model, consisting of a set of non-linear partial-differential equations – normally the Navier-Stokes equations – which are discretized both in space and time to result in a set of algebraic equations. We write these equations in the form

$$(1) \qquad X_{i+1} = X_i + \Delta t\, G_i(X_i),$$

where X_i is an N-dimensional vector describing the state of the model at time τ_i and where the operator G_i depends on the scheme used for the discretization. The specification of an initial condition X_0 at time τ_0 determines a unique solution to (1) for all i (to simplify the discussion we ignore the additional problems raised by the lateral boundary conditions).

We now wish to compare the experimental data with the fields predicted by the model. This is not straightforward because the parameters observed experimentally may differ in nature from those described by the model. For example, the model may describe temperatures, but some of the data may consist of radiances. Also in the gridpoint model, the observations may take place at points which do not belong to the grid. Therefore we first need to transform the model variables X_i, if necessary, so that they can be compared with the observations. Let us assume that this is done by means of the transformation

$$(2) \qquad \hat{X}_i = H_i(X_i),$$

where H_i is the observation operator associated with the observation vector $\mathbf{Z_i}$. In practice a spectral method is used, so that H_i also represents the transformation from spectral space to state space.

We now define the objective function to be the distance between the fields predicted by the model X_i and the observation vectors $\mathbf{Z_i}$

$$(3) \qquad J = \frac{1}{2}\sum_{i=1}^{n}(H_i(X_i) - \mathbf{Z_i})^T\, \mathbf{W}\, (H_i(X_i) - \mathbf{Z_i}).$$

The weighting matrix $\mathbf{W}$ should ideally be defined as the inverse of the error covariance matrix [25], but in practice it is a diagonal matrix that attempts to capture some of this information. A crucial point is that the gradient ∇J can be computed at approximately the same cost as the objective function J by using the adjoint method [22]. We can now state the optimization problem: Find an initial condition vector X_0 such that the sequence of states $X_1,...,X_n$ given by (1), minimizes the distance function J.

Let us consider the size and structure of this objective function. The number of elements N in X_0, i.e. the number of variables in the optimization problem, is typically of the order

10^6. The number of time steps n is of the order 10^2. The evaluation of the objective function J is very expensive since it requires a solution of the model. For example on a Cray-90 the computation of J and its gradient ∇J could take 20 minutes or more. Thus the four-dimensional variational assimilation approach leads to a very large unconstrained optimization problem with an expensive objective function. But there is another major difficulty: the transformations H_i in (2) make the Hessian of J essentially dense.

Since weather forecasts have to be produced several times a day, and since the time required for the function evaluations is so large, simulations are currently terminated when a prescribed number of function evaluations has been reached. Nonlinear optimization therefore plays a crucial role in weather forecasting: if we could design faster optimization methods for this problem, the resolution of the models can be increased resulting in better forecasts. Furthermore, to obtain improved medium-range forecasts (4 to 5 days) it is necessary to include more nonlinearities in the model, which makes the optimization more difficult.

Due to the characteristics of this optimization problem, many weather forecasting codes use the Fletcher-Reeves or Polak-Ribière conjugate gradient methods described in the next section. Recently limited memory methods have been incorporated in some codes and appear to be more successful. But both types of methods still require too many evaluations of the model (function and gradient) and faster algorithms need to be developed. These new algorithms must be effective for problems with the following characteristics: the objective function is very expensive to compute, the gradient can be obtained at approximately the same cost as the function but the Hessian matrix cannot be computed or stored; the dimension of the problem is so large $n \approx 10^6$ that only a few (between 20 and 50) n-vectors can be stored in high-speed memory. Let us keep these restrictions in mind during the discussion that follows.

3 Nonlinear Conjugate Gradient Methods.

Nonlinear conjugate gradient methods are well suited for large-scale problems due to the simplicity of their iteration and their very low memory requirements. We will now give an overview of these methods and describe a few open research questions.

The Fletcher-Reeves nonlinear conjugate gradient method is designed to solve the unconstrained optimization problem

$$\min_x f(x). \tag{4}$$

Here $f : \mathbf{R}^n \to \mathbf{R}$ is a smooth, nonlinear function whose gradient will be denoted by g. The easiest way to motivate the Fletcher-Reeves method is as an extension of the linear conjugate gradient method, which can be considered as an algorithm for minimizing convex quadratic functions of the form

$$\min_x \phi(x) = \tfrac{1}{2} x^T A x - b^T x. \tag{5}$$

Here A is a symmetric and positive definite $n \times n$ matrix and b is an n-vector. We now present the two algorithms – the linear and nonlinear conjugate gradient methods – side by side. Note that the residual

$$r(x) = Ax - b,$$

coincides with the gradient of the quadratic ϕ.

LINEAR CG		**NONLINEAR CG**
Problem: $\min \frac{1}{2}x^T Ax - b^T x$		Problem: $\min f(x)$
$\beta_k = \|r_k\|^2 / \|r_{k-1}\|^2$	(a)	$\beta_k = \|g_k\|^2 / \|g_{k-1}\|^2$
$d_k = -r_k + \beta_k d_{k-1}$	(b)	$d_k = -g_k + \beta_k d_{k-1}$
$\alpha_k = r_k^T r_k / d_k^T A d_k$	(c)	α_k : line search
$x_{k+1} = x_k + \alpha_k d_k$	(d)	$x_{k+1} = x_k + \alpha_k d_k$
$r_{k+1} = r_k + \alpha_k A d_k$	(e)	evaluate g_{k+1}

Making the correspondence $r \leftrightarrow g$ we see that the two algorithms differ only in steps (c) and (e); therefore the only significant difference is in the line search. Note that the nonlinear CG method does not require access to any matrices, and can be implemented using only 3 n-vectors of storage. Later we will describe how the line search is performed in the nonlinear conjugate gradient method; this turns out to be a complex and crucial ingredient of the method. It suffices to say that if the objective function is quadratic the exact one-dimensional minimizer is found, so that in the quadratic case the two algorithms are identical.

There is an alternative expression, proposed by Polak and Ribière, for the parameter β_k in the nonlinear CG method,

$$(6) \qquad \beta_k = \frac{g_k^T(g_k - g_{k-1})}{g_{k-1}^T g_{k-1}}.$$

Due to the orthogonality of gradients in the linear CG method, (6) is equivalent to expression given in the Fletcher-Reeves method when the objective function is quadratic. However, when applied to nonlinear functions, the Polak-Ribière formula (6) gives rise to a more effective algorithm; the reasons for this are understood in part [32].

Preconditioning is done in the same way in the linear and nonlinear CG algorithms, but its interpretation in the nonlinear case is more difficult because the Hessian matrix is not constant. One can consider either a constant preconditioner that attempts to improve the eigenvalue distribution of the Hessian matrix at the solution, or a varying preconditioner. How to develop preconditioners for the nonlinear CG method remains an important research question, which we discuss later on.

Let us contrast the role that the linear and nonlinear CG methods play in their respective fields of application.

The linear conjugate gradient method plays a central role in numerical linear algebra. As indicated by the contributions in this book, current research focuses on its extensions to indefinite and nonsymmetric problems, on preconditioning techniques and on the study of its numerical stability.

Nonlinear conjugate gradient methods are *not* among the fastest or more robust optimization algorithms available today, but they remain very popular for solving large problems. This popularity is due in part to their simplicity and due to the fact that they were the first methods that could be applied to very large problems in which the Hessian matrix is too costly to compute or store. The more recently developed limited memory methods [23],[17],[14],[38] are more predictable, robust and tend to require fewer function evaluations than nonlinear CG methods. However, even in their most economical implementations, limited memory methods can require at least three times the storage – an important consideration when storage is at premium. Because of this economy of storage, research in nonlinear conjugate gradient methods continues to this day.

3.1 Open Questions

The Polak-Ribière and Fletcher-Reeves methods have a fundamental weakness: the search directions tend to be poorly scaled and the line search must perform one or more function evaluations to obtain a suitable steplength α_k. It is common to see steplengths that differ from 1 by two orders of magnitude (they can be larger or smaller than 1 depending on how the problem is scaled). Moreover the size of α_k tends to vary in an unpredictable manner. This is in sharp contrast to quasi-Newton and limited memory methods which accept the unit steplength most of the time, and thus usually require only one function evaluation per search direction.

The efficiency of nonlinear conjugate gradient methods would greatly improve if one could design a variant of the Fletcher-Reeves or Polak-Ribière methods that produced well-scaled search directions without increasing the storage requirements of the iteration. This is an interesting research topic, that has received considerable attention, but in which little progress has been made. Numerous authors have suggested search directions of the form

$$(7) \qquad d_k = -H_k g_k + \beta_k d_{k-1}$$

where H_k is a simple symmetric and positive definite matrix often satisfying a secant equation [21]. However, if the definition of H_k requires several vectors, then the storage requirements are similar to those of limited memory methods and the performance is not as good.

This brings us to the question of preconditioning. Algorithms of this form, or variations of it, have been studied by numerous authors, but in our view have been unsuccessful. The drawback is that if H_k contains useful information about the Hessian of the objective function, we are better off using the search direction $d_k = -H_k g_k$ since the addition of the last term in (7) may prevent d_k from being a descent direction unless the line search is sufficiently accurate. At least this has been the case in our unsuccessful attempts to devise iterations of the form (7).

One question that has been considered for many years, but has never been decisively resolved is whether nonlinear conjugate gradient methods can be used effectively to solve inequality constrained problems. Consider for simplicity the bound constrained problem

$$\min_x f(x)$$

$$\text{subject to} \quad l \leq x \leq u,$$

where f is a quadratic function and l and u are n-vectors. Suppose that we begin from the interior of the feasible region and that we generate a sequence of conjugate directions that decrease the objective function at each step. Let us now suppose that the iterates encounter one of the bounds. A natural idea is to make this bound active, i.e. to fix the variable that reached its bound and continue minimizing with respect to the other variables. At this point, however, one normally restarts the conjugate gradient iteration along the steepest descent direction, since the information from the previous search directions is not useful (and can in fact be harmful). This type of algorithm will not be efficient if changes in the active set occur frequently, since too many steepest descent steps will be taken. A possible solution would be to take the view of interior-point methods and devise a nonlinear conjugate gradient iteration in which the bounds on the variables are not dealt with explicitly. To the best of my knowledge, no such methods have yet been proposed. One of the crucial questions would be what to do when the barrier parameter changes. At least it will change less often than the active bounds.

Another question that has intrigued me for some time is whether one can take advantage of problem structure to design a more effective nonlinear conjugate gradient iteration. In particular, when the problem is partially separable [8] we can use the idea of partitioned updating [19] to obtain a very powerful quasi-Newton method. This is because the information contained in the partially separable description of the function is so detailed that one can focus in exploring the objective function only along the relevant directions, i.e. one can ignore the invariant subspace [8] of the function and only consider its complement. An open question is whether one can use this type of invariant-subspace information to design nonlinear conjugate gradient methods.

An interesting new direction of research is the design of nonlinear conjugate gradient methods on geodesics [12].

3.2 Convergence Properties of Nonlinear CG Methods

It is possible to describe very succinctly some theoretical properties common to all nonlinear conjugate gradient methods. Let us consider any iteration of the form

$$(8) \quad d_k = -g_k + \beta_k d_{k-1}$$

$$(9) \quad x_{k+1} = x_k + \alpha_k d_k,$$

where α_k is a steplength satisfying the strong Wolfe conditions [32] and β_k is any scalar. A naive examination may suggest that the form of (8) is a recipe for disaster, since by adding all the gradients into the search directions these may grow without bound. It turns out that this simple observation is not totally misguided, since Powell has shown [36] that this could happen to the Polak-Ribière method. Moreover the growth in the length of the search direction is a meaningful quantity in the convergence analysis, as we now discuss.

Let us assume that the starting point x_0 for the iteration (8)-(9) is such that the level set $L = \{x : f(x) \leq f(x_0)\}$ is bounded, and that the objective function is twice continuously differentiable on L. Then one can show [36] that the iteration can fail, in the sense that

$$(10) \quad \|g_k\| \geq \gamma > 0 \quad \text{for all } k,$$

only if $\|d_k\| \to \infty$ sufficiently rapidly. To be more precise, the sequence of gradient norms $\|g_k\|$ can be bounded away from zero only if

$$(11) \quad \sum_{k \geq 0} \frac{1}{\|d_k\|} < \infty.$$

This observation can be used as the basis for the global convergence of nonlinear conjugate gradient methods. For example, one can show that the Fletcher-Reeves method

is globally convergent, using the following argument by contradiction. Suppose that (10) holds; then we can show that the search directions generated by the Fletcher-Reeves method satisfy
$$\|d_k\|^2 \le ck,$$
for some constant c, which implies that
$$\sum_{k \ge 0} \frac{1}{\|d_k\|} = \infty.$$

However this contradicts (11) which is a necessary condition for the algorithm to fail. Hence the assumption (10) must be false and we conclude that $\liminf_{k \to \infty} \|g_k\| = 0$. This means that the Fletcher-Reeves method is globally convergent to stationary points of the problem.

It is surprising that global convergence has not been established for the Polak-Ribière method using the standard Wolfe conditions. Nevertheless Grippo and Lucidi [20] describe new line search conditions that are designed to match the requirements of the convergence theory, and that ensure that the Polak-Ribière method is globally convergent. It is not yet known if the line search conditions of Grippo and Lucidi have some practical disadvantages, and their numerical performance is currently being investigated.

Another interesting result is that the Polak-Ribière method can be made globally convergent using standard Wolfe conditions if the parameter β_k in (8) is not allowed to be negative, i.e. if we define
$$\beta_k = \max\{\beta_k^{PR}, 0\}.$$

Most of these results are difficult to motivate or to explain intuitively, and I find the behavior of nonlinear conjugate gradient iterations to be quite complex and surprising.

4 Uses of the Linear CG Method in Nonlinear Optimization

Our discussion now moves from nonlinear CG methods to the role played by the (linear) CG iteration in two important optimization algorithms. The first is a truncated Newton method with trust regions in which some properties of the linear CG method are cleverly exploited to obtain global convergence as well as a fast asymptotic convergence rate. The second algorithm is a reduced Hessian method for constrained optimization; here the open question is how to design preconditioners for the CG iteration.

4.1 Truncated Newton Method

When the number of variables is large, it is not cost-effective to compute the Newton step accurately at every iteration. Instead it is attractive to approximately solve Newton equations

(12) $$\nabla^2 f(x_k) d = -g_k$$

by an iterative technique such as CG. It is also well known that Newton's method is only locally convergent, and to obtain convergence from remote starting points it is necessary to introduce either a line search or a trust region strategy [10], [13]. Steihaug [40] showed that we can achieve efficiency and global convergence by an elegant use of the CG method in a trust region framework. An interesting feature of this approach is that it can be applied regardless of whether the Hessian matrix $\nabla^2 f(x_k)$ is positive definite or not, as we now discuss.

At every iterate x_k we choose a trust region radius Δ_k and compute a step d_k by approximately solving the quadratic subproblem

(13) $$\min \phi(x) = g_k^T d + \tfrac{1}{2} d^T \nabla^2 f(x_k) d$$

(14) $$\text{subject to } \|d\|_2 \leq \Delta_k.$$

This approximate solution d_k is obtained by applying the CG method to (12) and stopping when one of several conditions is satisfied. To be precise, we choose the initial guess $d^0 = 0$ and apply the CG method to (12), to generate a sequence of approximate solutions $\{d^j\}$ and a sequence of conjugate directions $\{p_j\}$, and terminate when one of the following three conditions is satisfied:

(c1) The residual $r_j = \|\nabla^2 f(x_k) d^j + g_k\|$ is smaller than a prescribed tolerance t_k.

(c2) One of the approximate solutions d^{j+1} satisfies $\|d^{j+1}\| \geq \Delta_k$; in other words the CG iteration has stepped outside the trust region (14). In this case we simply backtrack to the trust region by finding $\tau > 0$ such that $\|d^j + \tau p_j\| = \Delta_k$, and define the approximate solution of the trust region problem (13)-(14) to be $d_k = d^j + \tau p_j$.

(c3) One of the conjugate search directions p_j generated by the CG iteration is a direction of negative curvature, i.e. $p_j^T \nabla^2 f(x_k) p_j < 0$. In this case we follow this direction of negative curvature to the boundary of the trust region. The approximate solution d_k of the trust region subproblem is thus defined as $d_k = d^j + \tau p_j$, where τ is defined so that $\|d_k\| = \Delta_k$.

It is crucial that the initial estimate d^0 in the CG iteration be $d^0 = 0$, for in this case one can show [40] that the approximate solutions increase in length at each step, i.e.

$$\|d^j\| < \|d^{j+1}\|,$$

for all j. Therefore if the CG iteration leaves the trust region, it will never return to it. The solution generated by Steihaug's strategy therefore gives the lowest value of the quadratic objective ϕ in the span of the CG directions and inside the trust region.

A remarkable property of this approach is that it guarantees that the optimization algorithm is globally convergent. This follows from the fact that the first step in the CG iteration is along the negative residual of (12), and since the initial guess is $d^0 = 0$, this residual equals the steepest descent direction $-g_k$ of the nonlinear objective function. Thus the first step of the CG process minimizes the quadratic model ϕ along the steepest descent direction – the resulting step is called the Cauchy step in the optimization literature – and all subsequent steps reduce the value of the ϕ. The convergence theory of trust region methods [35] states that a method is globally convergent if the step d_k gives a reduction of the model ϕ that is at least as good as that given by the Cauchy step. Thus global convergence of Steihaug's method follows in a straightforward way.

The algorithm can also be made rapidly convergent by tightening the convergence tolerance t_k in (c1). Conditions on t_k that guarantee superlinear or quadratic convergence are given in [11]. Since in a neighborhood of a strong minimizer the Hessian matrix $\nabla^2 f(x_k)$ is positive definite, the CG iteration will not encounter negative curvature. Moreover the strategy for adjusting the trust region also guarantees that, as the iterates approach the solution, the trust region becomes inactive in the sense that all the approximate steps generated by the CG iteration lie inside the trust region regardless of how tight the convergence tolerance is. In conclusion the components of the algorithm fit in nicely to give all the desired properties: global convergence, fast asymptotic rate and economy of implementation.

Steihaug's method can also be used in algorithms for constrained optimization, and it is now being incorporated into a variety of optimization codes. Because of its increased importance, it is receiving closer scrutiny and this has given rise to several open questions. The first has to do with the fact that, when the CG process detects negative curvature, little is known about the quality of this direction. It could correspond to a negative eigenvalue with small magnitude, and much more progress could be made by seeking a direction corresponding to a more negative eigenvalue. In other words, should there be criteria of what a good direction of negative curvature is? There is also the possibility that CG is unstable on indefinite problems and could produce erratic results; can this be a problem in practice? Recent proposals to improve upon Steihaug's method are given in [2] and [1], but it is too early to say if they give rise to improvements in practice. An alternative is to seek to solve the trust region problem (13)-(14) exactly, and a new method for doing this in the large-scale case has recently been proposed in [39].

4.2 Solving Reduced Hessian Systems

An important open question concerning the use of the conjugate gradient method arises in constrained optimization. Here the problem is

(15)
$$\min\ f(x)$$
$$\text{s.t.}\ c(x) \leq 0,$$

where $f(x) : \mathbf{R}^n \to \mathbf{R}$ and $c(x) : \mathbf{R}^n \to \mathbf{R}^t$ are smooth, nonlinear functions. We are interested in the case when the number of variables n is in the hundreds or thousands, but the number of constraints t can be large or small.

One of the most successful approaches for solving (15) is the sequential quadratic programming (SQP) method [18],[13] in which the constraints of the problem are linearized and the objective function is replaced by a quadratic approximation of the Lagrangian of (15). The SQP method can be implemented in various ways; in one of them the linearized constraints are eliminated and the step of the algorithm is obtained by solving a reduced Hessian system of the form

(16)
$$Z_k^T W_k Z_k u = -Z_k^T \bar{g}_k.$$

Here Z_k is a basis for the tangent space of the constraints, W_k is the Hessian of the Lagrangian or an approximation to it – and is assumed to be positive definite – and $Z_k^T \bar{g}_k$ is called the reduced gradient. The linear system (16) can be solved by the CG method, but trying to precondition CG seems to be very difficult because the matrix Z_k is not known explicitly, as we now explain.

The SQP method selects a subset of the constraints of the problem (say the first p) and these constraints are considered active [18]. If we let $A_k = [\nabla c^1(x_k) \cdots \nabla c^p(x_k)]$ denote the matrix of gradients of these constraints, then Z_k is defined as a full rank $n \times t$ matrix such that $A_k^T Z_k = 0$. Ideally Z_k should have orthonormal columns to prevent numerical instabilities, but this can be prohibitively expensive for large problems. In practice Z_k is chosen to represent a simple elimination of variables. We partition A_k^T as

$$A_k^T = [C_k\ N_k],$$

where the $t \times t$ basis matrix C_k is nonsingular, and then define

$$(17) \qquad Z_k = \begin{bmatrix} -C_k^{-1} N_k \\ I \end{bmatrix}.$$

We compute sparse LU factors for the basis matrix C_k, thereby obtaining an implicit representation of Z_k. The system (16) is solved by CG which requires only products of the matrix $Z_k^T W_k Z_k$ (which is never formed) with vectors $v \in \mathbf{R}^{n-t}$. Thus each product of Z_k or Z_k^T with a vector requires solving two triangular systems of size t. An important open question is how to precondition the CG iteration, given that the matrix in (16) is not known explicitly and that extracting information from it is expensive; for example simply computing its diagonal (which may not be a good preconditioner anyway) is costly. Some attempts to develop preconditioners are given in [28] and [2], but they may not be very helpful in practice. The question of preconditioning is so important that if it cannot be resolved, this reduced Hessian approach will not give rise to robust general-purpose codes.

5 New Algorithms

We will now consider some new algorithmic ideas that attempt to more fully exploit the interrelationship between the linear CG method and nonlinear optimization algorithms. Throughout this section we restrict our attention to the unconstrained optimization problem (4).

The recently developed *iterated subspace minimization method* [9] is based on the observation that the step computation in Newton-type methods can be quite expensive for large problems, and it may therefore be advantageous to use the information generated during the iteration more fully, even if this results in additional function evaluations. For example, an algorithm studied in [9] is based on a truncated Newton method in which the step is computed by an inner CG iteration, as described in the previous section. However, instead of accepting the truncated Newton step, one selects a few of the directions $\{p_j\}$ generated during the inner CG iteration, and uses them to construct a subspace S_k over which the nonlinear objective function is minimized further. The subspace S_k is defined as the span of (some or all) the directions generated during the inner CG cycle, and the nonlinear objective function f is approximately minimized over S_k. Since the dimension of the subspace S_k is small, this is a low-dimensional nonlinear optimization problem that can be solved, for example, by means of the BFGS method. The basis of this approach is that CG is likely to determine important directions over which the nonlinear objective function needs to be explored further. A more detailed description of this approach is given in the contribution by Conn et al in this volume.

We have recently developed [5] a new algorithm that takes a different point of view. It is designed for problems like the one described in §2, where the Hessian matrix cannot be computed and where the cost of evaluating the function dominates the cost of the iteration. The goal is to design an algorithm with faster convergence rate than nonlinear conjugate gradient or limited memory methods, and that achieves this speedup without a significant increase in the number of function or gradient evaluations. We call this algorithm the Discrete Newton Method with Memory (DINEMO), and will now give a broad outline of it.

The truncated Newton method described in the previous section can be adapted to the case where the Hessian $\nabla^2 f(x_k)$ is not available, by noting that the inner CG iteration only requires the product of $\nabla^2 f(x_k)$ with the inner displacement vectors $\{p_j\}$ – and does not require the Hessian matrix itself. These products can be approximated by finite differences,

$$\nabla^2 f(x_k) p_j \approx \frac{g(x_k + \epsilon p_j) - g(x_k)}{\epsilon}, \tag{18}$$

where ϵ is a small differencing parameter. There is however a big price to pay for doing this: since each iteration of the inner CG method performs one product $\nabla^2 f(x_k) p_j$, it requires a new evaluation of the gradient g of the objective function. Thus one step of this *discrete Newton method* will require as many gradient evaluations as the number of iterations in the inner CG cycle [33],[27], [26], [37].

Even though the discrete Newton method converges much faster than nonlinear CG and limited memory methods, it has the drawback that the large amount of functional information spent in each major iteration is not saved and cannot be used in subsequent iterations. In contrast, quasi-Newton methods are very good at updating and preserving information gathered about the Hessian of the objective function. The key idea of the new algorithm is to introduce memory into the discrete Newton method, and to do so in the form of a limited memory quasi-Newton matrix. The curvature information contained in the gradient differences (18) is saved in the form of a limited memory matrix B_k, and if this matrix is judged to be of good quality – which will usually be the case – the next step of the algorithm will be a limited memory iteration which is quite inexpensive and requires no additional function evaluations.

Therefore the new algorithm combines discrete Newton and limited memory steps, but instead of simply alternating these two iterations, the information is merged to produce a new kind of algorithm. One can view this from a different perspective: the new algorithm can be seen as an attempt to improve the limited memory method by incorporating the curvature information generated during the inner CG iteration into the limited memory matrix B_k. From this perspective the key observation is similar to that made concerning the iterated subspace minimization method, namely that the inner CG iteration explores the objective function along some important directions that may never be considered by the pure limited memory method. In particular, the inner CG iteration is capable of collecting information along directions corresponding to small eigenvalues of the Hessian matrix; these directions are crucial when solving ill-conditioned problems and tend to be ignored by limited memory methods (as well as by quasi-Newton methods when the dimension of the problem is large).

Let us illustrate how the new algorithm operates. Suppose that at the iterate x_k we perform a discrete Newton step: we approximately solve (12) by the conjugate gradient method, using the finite differences (18) to avoid access to the Hessian. This CG iteration can be written as

$$d^{i+1} = d^i + \sigma^i p_i, \tag{19}$$

where σ^i is the steplength and the directions $\{p_i\}$ are conjugate with respect to the Hessian matrix $\nabla^2 f(x_k)$. The inner CG iteration is performed as indicated in the previous section, with one exception: if negative curvature is detected, the iteration is terminated and the estimate d^j found *prior* to detecting negative curvature is taken as the step of the optimization method. This is necessary because the information gathered during the inner CG iteration will be used to define a positive definite limited memory matrix, as follows.

Let m denote the number of CG iterations performed and define the vectors

$$s_k^i = \epsilon p_i, \qquad y_k^i = g(x_k + \epsilon p_i) - g(x_k), \quad i = 1, \ldots, m. \tag{20}$$

Since these vectors satisfy the curvature condition $(s_k^i)^T y_k^i > 0$, because the inner CG iteration is stopped before negative curvature is detected, they can be used to define a limited memory BFGS matrix B_{k+1} that approximates the Hessian of the objective function. We will not review here the way in which limited memory matrices are constructed [23],[4], but only mention that B_{k+1} is obtained by updating a multiple of the identity matrix $m+1$ times using the m pairs $(s_k^1, y_k^1), (s_k^2, y_k^2), ...$, as well as the pair (s_k, y_k) corresponding to the total step from x_k to x_{k+1}, which is defined as

$$s_k = x_{k+1} - x_k, \quad y_k = g_{k+1} - g_k.$$

(Note that difference between s_k and s_k^i). Having reached the new iterate x_{k+1}, we proceed with the limited memory method generating iterates $x_{k+2}, x_{k+3}, ...$, and updating the limited memory matrices $B_{k+2}, B_{k+3}, ...$, at each iteration by adding the information $(s_{k+1}, y_{k+1}), (s_{k+2}, y_{k+2}), ...$, from the total steps.

If at some iterate x_t we believe that a new discrete Newton step must be computed, we approximately solve the linear system

(21) $$\nabla^2 f(x_t) d_t = -g(x_t)$$

using the CG method as before, and storing the new inner correction vectors (20). This produces the search direction d_t leading to the new iterate x_{t+1}. At this point we remove from the limited memory matrix B_t the information obtained from the previous cycle of conjugate gradient iterations and replace it with the latest set of vectors (20) generated by the inner CG iterations used to solve (21).

Therefore at every iteration of the algorithm (except for the first one) the limited memory matrix contains two kinds of information: one corresponding to the pairs $\{s^i, y^i\}$ generated during the latest inner CG iteration, and the other corresponding to the pairs $\{s_k, y_k\}$ generate at each outer (or major) iteration of the algorithm. The information corresponding to the outer steps is continuously updated as in the standard limited memory method: the latest correction vector replaces the oldest one. The information from the inner CG iteration is kept intact until a new truncated Newton step is performed, at which point it is refreshed.

There are many variations of the algorithm just described. Some of these and numerical experiments are given in [5]. Computational experience shows that this approach can give excellent results, but that success depends on a wise termination condition for the inner CG iteration. How to design this stopping condition, and adapt it as the algorithm moves, remains an important open question of this research.

The iterated subspace minimization approach and DINEMO are examples of optimization research that attempts to devise more efficient algorithms for large-scale problems by fusing the properties of the CG and quasi-Newton methods.

6 Conclusion

There are interesting open questions concerning the design and implementation of nonlinear conjugate gradient methods. Advances in this area would be very useful in applications where storage is at a premium.

The linear CG method is playing an increasingly important role in algorithms for both unconstrained and constrained optimization. This is not only due to the fact that CG is a powerful iterative solver, but also because its properties merge very well with those of the nonlinear optimization methods. Much research needs to be done on Steihaug's strategy and on preconditioning. Finally, the interrelationship between quasi-Newton and conjugate gradient methods does not appear to have been fully exploited in the design of algorithms for large-scale problems, and some promising ideas that attempt to do so have recently been proposed.

References

[1] M. Arioli, T. F. Chan, I. S. Duff, N. I. M. Gould, and J. K. Reid, *Computing a search direction for large-scale linearly constrained nonlinear optimization calculations*, Technical Report TR/PA/93/94, CERFACS Toulouse, France, 1993.

[2] M.A. Branch, T.F. Coleman and Y.Li, *A Subspace, Interior, and Conjugate Gradient Method for Large-scale Bound-constrained Minimization Problems*, Tech. Report CTC95TR217, Cornell Theory Center, Cornell University, 1995.

[3] A. Buckley, *A combined conjugate gradient quasi-Newton minimization Algorithm*, Mathematical Programming, 15 (1978), pp. 200–210.

[4] R. Byrd, J. Nocedal and R. Schnabel *Representations of quasi-Newton matrices and their use in limited memory methods*, Math. Prog. Ser. A, Vol 63 (1994), pp. 129–156.

[5] R.H. Byrd, J. Nocedal and C. Zhu, *Towards a discrete Newton method with memory for large-scale optimization*, Optimization Technology Center Report OTC-95-1, EECS Department, Northwestern University, 1995.

[6] W.C. Davidon, *Variable metric methods for minimization*, Argonne National Lab Report, Argonne, IL., 1959.

[7] J.C. Derber, *Variational Four Dimensional Analysis Using Quasi-Geostrophic Constraints*, Mon. Wea. Rev. 115 (1987), pp. 998–1008.

[8] A.R. Conn, N.I.M. Gould, Ph.L. Toint, LANCELOT: *a FORTRAN package for large-scale nonlinear optimization (Release A)*, Number 17 in Springer Series in Computational Mathematics, Springer-Verlag, New York, 1992.

[9] A.R. Conn, N.I.M. Gould, A. Sartenaer and Ph.L. Toint, *On Iterated-Subspace Minimization Methods for Nonlinear Optimization*, IBM T. J. Watson Research Center Research Report RC 19578, 1994.

[10] J. E. Dennis, Jr. and R. B. Schnabel, *Numerical Methods for Unconstrained Optimization and Nonlinear Equations*, Prentice Hall, Englewood Cliffs, N.J., 1983.

[11] R.S. Dembo, S.C. Eisenstat, and T. Steihaug, *Inexact Newton methods*, SIAM J. Numer. Anal., 19 (1982), pp. 400–408.

[12] A. Edleman, T. Arias and S.T. Smith, *Conjugate gradient on the Stiefel and Grassmann Manifolds*, Department of Mathematics, Massachusetts Institute of Technology, 1995.

[13] R. Fletcher, *Practical Methods of Optimization*, John Wiley and Sons (second edition), Chichester, 1987.

[14] R. Fletcher, *Low storage methods for unconstrained optimization*, Computational Solutions of Nonlinear Systems of Equations, eds. E. L. Allgower and K. Georg, Lectures in Applied Mathematics, 26 1990, AMS Publications, Providence, RI.

[15] R. Fletcher and M.J.D. Powell, *A rapidly convergent descent method for minimization*, Comput. J., 6 (1963), pp. 163–168.

[16] R. Fletcher and C.M. Reeves, *Function minimization by conjugate gradients*, Computer Journal, 7 (1964), pp. 149–154.

[17] J.C. Gilbert and C. Lemaréchal, *Some numerical experiments with variable storage quasi-Newton algorithms*, Mathematical Programming, 45 (1989), pp. 407–436.

[18] P. Gill, W. Murray and M.H. Wright, *Practical Optimization*, Academic Press, 1981.

[19] A. Griewank and Ph.L. Toint, *On the unconstrained optimization of partially separable objective functions*, in Nonlinear Optimization 1981, (M.J.D. Powell, ed.), Academic Press, London, pp. 301–312, 1982a.

[20] L. Grippo and S. Lucidi, *A globally convergent version of the Polak-Ribière gradient method*, Dipartimento di Informatica e Sistematica, Universita degli Studi di Roma "La Sapeinza", R. 08-95, 1995.

[21] Y.F. Hu and C. Storey, *On unconstrained conjugate gradient optimization methods and their interrelationships*, Mathematics Report Number A129, Loughborough University of Technology, 1990.

[22] F.X. Le Dimet and O. Talagrand, *Variational Algorithms for Analysis and Assimilation of Meteorological Observations: Theoretical Aspects*, Tellus, 38A (1986), pp. 97–110.

[23] D. C. Liu and J. Nocedal, *On the limited memory BFGS method for large scale optimization methods*, Mathematical Programming, 45 (1989), pp. 503-528.

[24] R.B. Long and W.C. Thacker, *Data Assimilation Into a Numerical Equatorial Ocean Model, part 2: Assimilation Experiments*, Dyn. Atmos. Oceans, 13 (1989), pp. 465–477.

[25] A.C. Lorenc, *Analysis Methods for Numerical Weather Prediction*, Quart. J. Roy. Meteor. Soc., 112 (1986), pp. 1177-1194.

[26] S.G. Nash, *User's guide for TN/TNBC: FORTRAN routines for nonlinear optimization*, Report 397, Mathematical Sciences Dept., The Johns Hopkins University, 1984.

[27] S.G. Nash, *Preconditioning of truncated-Newton methods*, SIAM Journal on Scientific and Statistical Computing, 6 (1985), pp. 599–616.

[28] S. Nash and A. Sofer, *Preconditioning of Reduced Matrices*, Technical Report 93-01, Department of Operations, George Mason University, 1993.

[29] J.L. Nazareth, *The method of successive affine reduction for nonlinear minimization*, Mathematical Programming, 35 (1986), pp. 373-387.

[30] L. Nazareth, *A relationship between the BFGS and conjugate gradient algorithms and its implications for new algorithms*, SIAM Journal on Numerical Analysis, 16 (1979), pp. 794-800.

[31] J. Nocedal, *Updating quasi-Newton matrices with limited storage*, Mathematics of Computation, 35 (1980), pp. 773-782.

[32] J. Nocedal, *Theory of Algorithms for Unconstrained Optimization*, Acta Numerica, Cambdridge University Press, 1992.

[33] D.P. O'Leary, *A discrete Newton algorithm for minimizing a function of many variables*, Mathematical Programming, 23 (1982), pp. 20–33.

[34] A. Perry, *A class of conjugate gradient algorithms with a two-step variable metric memory*, Discussion paper no. 269, Center for Mathematical Studies in the Economics and Management Science, Northwestern University, 1977.

[35] M.J.D. Powell, *On the global convergence of trust region algorithm for unconstrained optimization*, Mathematical Programming, 29 (1984), pp. 297-303.

[36] M.J.D. Powell, *Nonconvex minimization calculations and the conjugate gradient method*, in Lecture Notes in Mathematics, 1066 (1984a), Springer-Berlag, Berlin, pp. 122–141.

[37] T. Schlick and A. Fogelson, *TNPACK - A truncated Newton package for large-scale problems: I. Algorithms and usage*, ACM Transactions on Mathematical Software, 18 (1992), no. 1, pp. 46-70.

[38] D. Siegel, *Implementing and modifying Broyden class updates for large scale optimization*, Report DAMTP 1992/NA12, Department of Applied Mathematics and Theoretical Physics, University of Cambridge, Cambridge, England, 1992.

[39] D.C. Sorensen, *Minimization of a large scale quadratic function subject to an ellipsoidal constraint*, Department of Computational and Applied Mathematics, Rice university, 1995.

[40] T. Steihaug, *The conjugate gradient method and trust regions in large scale optimization*, SIAM J. Num. Anal., 20 (1983), pp. 626–637.

Chapter 3
On Conjugate Gradient-Like Methods for Eigen-Like Problems*

Alan Edelman[†]   Steven T. Smith[‡]

Abstract

Numerical analysts, physicists, and signal processing engineers have proposed algorithms that might be called conjugate gradient for problems associated with the computation of eigenvalues. There are many variations, mostly one eigenvalue at a time though sometimes block algorithms are proposed. Is there a correct "conjugate gradient" algorithm for the eigenvalue problem? How are the algorithms related amongst themselves and with other related algorithms such as Lanczos, the Newton method, and the Rayleigh quotient?

1 Introduction

If we minimize $y^T A y$ on the unit sphere, perhaps with conjugate gradient optimization, we compute the smallest eigenvalue of A (assumed to be symmetric). Our objective function is quadratic, so we obtain the exact eigenvalue after n steps of conjugate gradient. The computation is mathematically equivalent to the Lanczos procedure. At the kth step, we obtain the optimal answer in a k dimensional Krylov space.

The above paragraph may sound correct, but it is wrong! The objective function is quadratic, but only because of the constraint. The Rayleigh quotient $r(y) = y^T A y / y^T y$ can be minimized without constraint, but it is not quadratic. The conjugate gradient link with the Lanczos process here is also dubious. How tempting it is to hear "eigenvalue" juxtaposed with "conjugate gradient" and instantly, almost by reflex, yell "Lanczos." To add to the confusion, conjugate gradient optimization of the Rayleigh quotient does compute an answer in a Krylov space, but not the optimal answer.

Our purpose in this paper is to 1) dispel a common myth, 2) create an intellectual framework tying together the plethora of conjugate gradient algorithms, 3) provide a scholarly review of the literature, and 4) show how the differential geometry point of view introduces new algorithms in a context that allows intellectual ties to older algorithms. We believe each of these purposes is justified, but the most important goal is to show that the differential geometry point of view gives idealized algorithms that unify the subject. Therefore in Section 1 we present the myth. In Sections 2 and 3 we dispel the myth and

---

*This paper will also appear in the journal *BIT*.

[†]Department of Mathematics Room 2-380, Massachusetts Institute of Technology, Cambridge, MA 02139, edelman@math.mit.edu, Supported by a fellowship from the Alfred P. Sloan Foundation and NSF Grant 9404326-CCR.

[‡]M.I.T. Lincoln Laboratory, 244 Wood Street, Lexington, MA 02173-9108, stsmith@ll.mit.edu. This work was sponsored by ARPA under Air Force contract F19628-95-C-0002. Opinions, interpretations, conclusions, and recommendations are those of the author and are not necessarily endorsed by the United States Air Force.

identify the common features and design choices for a large class of algorithms. Section 4 reviews the literature in the context of this framework. Section 5 discusses the differential geometry methods in this framework. Finally Section 6 presents an entirely new perspective on Newton's method and discusses connections to previous work by Chatelin, Demmel, and others.

2 Conjugate Gradient: linear, nonlinear, and idealized

Our first task is to eliminate any confusion among the various conjugate gradient algorithms with the Lanczos algorithm. This confusion has appeared in the literature. We hope that a naming convention will help distinguish the algorithms in the future:

CONJUGATE GRADIENT ALGORITHMS

LCG **Linear Conjugate Gradient:**
Standard method for solving $Ax = b$, where A is a symmetric positive definite matrix. Found in all numerical linear algebra references. May be viewed as minimizing the objective function $f(x) = \frac{1}{2}x^T A x - x^T b$.

NCG **Nonlinear Conjugate Gradient:**
A well-known algorithm often used for unconstrained minimization. Found in all numerical optimization texts. May be used to minimize very general objective functions. Each step in the algorithm requires the solution to a one dimensional line minimization problem.

ICG **Idealized Conjugate Gradient:**
A fictional algorithm introduced here for purposes of exposition. We define ICG as an algorithm whose kth iterate x_k satisfies $f(x_k) = \min_{x \in \mathcal{K}_k} f(x)$, where $\mathcal{K}_k$ denotes a k dimensional Krylov space. The Krylov spaces are nested.

The three conjugate gradient algorithms are equivalent for the special case of quadratic objective functions with positive definite Hessians. To say this in a slightly different way, the LCG algorithm, already a special case of the more general NCG algorithm, may be derived with the "one line minimization per iteration" point of view, and nevertheless, the algorithm has the "global k-dimensional" property of an ICG anyway.

To contrast, if the function is not quadratic with positive definite Hessian, there is no expectation that the kth step of NCG will have the ICG property of being the global minimum in a k-dimensional space. The best that we hope for is that the ICG property holds approximately as we approach the minimum.

Similar to ICG is s-step steepest descent. This algorithm minimizes an objective function in the s-dimensional space spanned by the most recent s gradients. If the gradients fall in a certain natural Krylov space, then there is no distinction between ICG and s-step steepest descent.

The word *conjugacy* merits discussion. For LCG, all search directions are all conjugate with respect to the Hessian matrix A, i.e., $p_i^T A p_j = 0$, $i \neq j$. For NCG, consecutive directions are conjugate with respect to the function's Hessian matrix; Fletcher-Reeves and Polak-Ribière approximate this conjugacy. Generally, ICG search directions are not conjugate; perhaps a different name is appropriate. However, ICG can be viewed as an extension of LCG.

Block versions of NCG and ICG may be defined. A block NCG replaces a 1-dimensional line search with a p-dimensional line search. A block ICG maximizes over block Krylov spaces of size p, $2p$, $3p$, etc.

Consider the case when the objective function is the Rayleigh quotient, $f(y) = (y^T A y)/(y^T y)$.

- An NCG algorithm for Rayleigh quotient optimization computes an answer in a Krylov space. This answer is not the optimal choice from that subspace. Therefore the NCG algorithm is not an ICG algorithm. The exact eigenvalue will not be obtained in n steps. Such an algorithm is not equivalent to Lanczos.

- The Lanczos algorithm (including the computation of the tridiagonal's smallest eigenvalue) is an ICG algorithm for the Rayleigh quotient. Lanczos does compute the exact eigenvalue in n steps. This algorithm is not an NCG algorithm because it is not equivalent to an algorithm that performs line searches.

- There is also the LCG link with Lanczos through tridiagonalization. The residual vectors computed by conjugate gradient are (up to normalization) Lanczos vectors that tridiagonalize the matrix. The link is through the tridiagonalization and not the eigenvalue computation. There is no link to the NCG algorithm for eigenvalue computation.

Also we note that

- any algorithm that requires that the symmetric matrix A be positive definite is not a correct NCG algorithm for the Rayleigh quotient. Since all derivatives of the objective function are invariant under shifts, so should be the algorithm.

Understanding this last point clearly is the key sign that the reader understands the distinction between NCG for the Rayleigh quotient and LCG which is an optimization on quadratic functions. LCG requires positive definite quadratic optimization functions so that a minimum exists. The Rayleigh quotient always has a minimum for symmetric A; no positive definite condition is needed.

The first choice in our design space for algorithms is to consider whether to include the constraints or work with the Rayleigh quotient. The second choice is whether to have a one eigenvalue algorithm or a block algorithm. We indicate these choices below:

RAYLEIGH QUOTIENTS
Unconstrained one eigenvalue Rayleigh quotient $r(y) = (y^T A y)/(y^T y)$
Constrained Rayleigh quotient $r(y) = y^T A y \ (y^T y = 1)$

Unconstrained block Rayleigh quotient $R(Y) = \text{tr}(Y^T A Y)(Y^T Y)^{-1}$
Constrained block Rayleigh quotient $R(Y) = \text{tr} Y^T A Y \ (Y^T Y = I_p)$

In the above box Y denotes an $n \times p$ matrix. Of the four choices, only the constrained $r(y) = y^T A y$ is (generically) non-degenerate in that the global minima are points in n dimensions, as opposed to lines or subspaces.

The third choice in our design space for an NCG eigenvalue algorithm is how to pick the new search direction. The various options are

NCG SEARCH DIRECTION CHOICES	
FR	The Fletcher-Reeves approach
PR	The Polak-Ribière approach
CA	Conjugacy through the matrix A
CSH	Conjugacy through the singular free space Hessian H
CCH	Conjugacy through a non-singular constrained Hessian H

All of these choices, except for CA, are reasonable. There is no sound mathematical justification for CA, though it has been proposed in the literature.

3 Comparison between NCG and Lanczos

Since Lanczos is an ICG algorithm for the Rayleigh quotient, why would anyone consider the construction of an NCG algorithm which is guaranteed to give a worse answer for the same number of matrix vector multiplies? Ignoring the extra storage that may be needed in the Lanczos algorithm, the more practical point is that in many algorithms it is not the eigenvalues of a matrix that are desired at all.

In the applications of interest, a matrix may be changing in "space" or time. What we want from an algorithm is to be able to track the changes. In many such applications, the problem sizes are huge so computational efficiency is of utmost importance. We have two particular examples in mind. In the Local Density Approximation (LDA) to Schrödinger's equation, the function to be minimized is not the Rayleigh quotient at all, but rather the LDA energy. In many applications, the energy function has the same degeneracy as the Rayleigh quotient in that it depends only on the span of the columns of Y rather than all elements of Y. Lanczos would not directly apply unless one wants to pretend that the function really is the Rayleigh quotient for a few iterations. Such pretense has lead to instabilities in the past.

Another class of problems arise in signal processing where the matrix is changing in time. Once again, Lanczos does not directly apply, unless one wants to pretend that the matrix is constant for a few intervals.

4 A History of CG for eigen-like problems

The purpose of this section is to demonstrate the need for a unifying theory of conjugate gradient algorithms for the eigenproblem by giving an annotated chronology of the references. We note that algorithms have been proposed by researchers in a number of disciplines, so it seems likely that researchers in one field may well be unaware of those from another.

In order to keep the chronology focused, we only included papers that linked conjugate gradient to eigenvalue or eigenvalue-like problems. We found it convenient to divide the papers into five broad categories:

- **Single Eigenvalue (NCG) Algorithms** This refers to the one eigenvalue at a time algorithm that was originally proposed by Bradbury and Fletcher [4]. This algorithm minimizes $r(y)$. Other eigenvalues may be obtained by deflation. The historical development covers a number of points in the design space.

- **Block (NCG) Algorithms** These are attempts at computing multiple eigenvalues simultaneously using a conjugate gradient style algorithm. The objective function is $R(Y)$. It is our point of view that none of these algorithms are a true conjugate gradient algorithm, though many important features of a true algorithm may be recognized in either explicit or rudimentary form in the proposed algorithms. The block case exhibits difficulties not found in the single eigenvalue case, because not only is there the orthogonality constraint $Y^T Y = I$, but also there is the "Grassmann equivalence" $R(Y) = R(YQ)$ for orthogonal matrices Q. This degeneracy may well be overcome without differential geometry, but one effective and beautiful theoretical approach towards working with these equivalence classes is the total space/base space viewpoint from differential geometry.

- **ICG: The Lanczos Link** Here our goal was not to mention papers that were connected to the Lanczos algorithm, but papers that looked at the Lanczos algorithm as an optimization algorithm with something of a conjugate gradient flavor. As mentioned earlier, the famous link between Lanczos and LCG is of no relevance here and is not included.

- **Differential Geometry Viewpoint** These papers make the link between numerical linear algebra, optimization, and differential geometry. They give a new theoretical viewpoint on the algorithms discussed in the other areas. This viewpoints shows that constrained algorithms may be implemented without explicit constraints. Ultimately, they answer the question of what it *means* to do conjugate gradient minimization for the eigenvalue problem.

- **Application I: LDA Schrödinger's Equation** Rightly or wrongly, conjugate gradient has become an extremely popular new method for scientists working with the local density approximation to Schrödinger's Equation. Here the problem is akin to minimizing the block Rayleigh quotient, but it is in fact more complicated. Many researchers are currently trying variations on the same theme. We only need to list the few that we were most aware of. One point of view is that they are computing eigenvalues of a matrix that is in some sense varying with space.

- **Application II: Adaptive Signal Processing** Here the algorithm is an eigenvalue algorithm, but the matrix may be thought of as changing with time. The goal is to do subspace tracking. All the papers in this category have already been listed in a previous category, but we thought it worthwhile to collect these papers under one heading as well.

Chronology of Algorithms

Key to search direction choices: FR=Fletcher-Reeves, PR=Polak-Ribière, CA=Conjugacy through A, CSH=Conjugacy through singular Hessian H, CCH= Conjugacy through constrained Hessian H.

1. **Single Eigenvalue Algorithms**

1966	Bradbury and Fletcher [4]	Notes degeneracy of Rayleigh quotient. Proposes projection to ∞- and 2-norm unit spheres. FR.
1969	Fox and Kapoor [15]	Application in Structural Dynamics. Downplays importance of the constraint. FR.
1969	Fried [16]	Application in Structural Dynamics. FR.
1971	Andersson [2]	Compares two norms mentioned above [4].
1971	Geradin [20]	Unconstrained CSH.
1972	Fried [17]	Unconstrained FR.
1974	Ruhe [30]	Systematically compares above CG algorithms with other non-CG algorithms. For CG, prefers unconstrained approach in [15] and [20]. Observes spectral influence on convergence.
1984	Fuhrmann and Liu [19]	Precursor to intrinsic geometry: Searches along geodesics. FR.
1986	Perdon and Gambolati [29]	CA with a proposed preconditioner.
1986	Chen, Sarkar, et al. [7]	Rederives known algorithms. CA.
1987	Haimi-Cohen and Cohen [23]	Rederives results such as those in [4].
1989	Yang, Sarkar, and Arvas [39]	Considers effect of normalization. Concludes that their 1986 proposal is not competitive with CSH or other choices. FR or PR style. [38].

2. Block Algorithms:

1971	Alsén [1]	Assumes A positive definite. Exact line maximization (not minimization) replaced by one step of orthogonal iteration. Corresponding minimization algorithm would require orthogonal inverse iteration. FR.
1982	Sameh and Wisniewski [31]	Not exactly a CG method but very similar. Constrained objective function with unconstrained block line searches. Section 2.3 suggests block line minimization on unconstrained function as a least squares problem to be solved with LCG. Makes the Lanczos style link through simultaneous iteration.
1995	Fu and Dowling [18]	Optimization of unconstrained $R(Y)$. Block search directions arranged column by column. Projects back to constraint surface with Gram Schmidt. CA.

3. ICG: The Lanczos Link:

1951	Karush [24]	An ICG algorithm (with restarts) for the Rayleigh quotient. Not explicitly identified with Lanczos or CG.
1974	Cullum and Donath [9, 10]	Identifies block Lanczos as a block ICG.
1978	Cullum [8]	Shows that block NCG on the Rayleigh quotient computes a (generally non-optimal) answer in the block Krylov space over which the ICG would finds the minimum.
1985	Cullum and Willoughby [11]	Chapter 7 summarizes optimization interpretation. CG used at times to mean NCG and other times ICG.

4. Differential Geometry Approach:

1993	Smith [32, 33]	Introduces Differential Geometry viewpoint. Hessian is replaced with second covariant derivative. Line searches replaced with geodesic searches.
1995	Edelman, Arias, Smith [14]	Works out details of the Grassmann manifold and Stiefel manifold approaches, including the total space-base space formulation, parallel transport, differing metrics, and the linear algebra links.

5. LDA Schrödinger's Equation

1989	Gillan [21]	Projects onto constraint space, search in tangent space, suggests NCG preconditioning, FR.
1989	Štich, Car, et al. [35]	Minimizes Rayleigh quotient only (not LDA energy) for simplicity.
1989	Teter, Payne, and Allan [37]	
1992	Arias [3]	Forms unconstrained analogue of Rayleigh quotient for LDA.
1992	Payne, Teter, Allan et al. [28]	Survey of minimization in LDA.
1993	Kresse and Hafner [25]	Molecular dynamics of liquid metals.
1994	Sung, Kawai, and Weare [36]	Computes structure of Lithium.

6. Adaptive Signal Processing

1986	Chen, Sarkar, et al. [7]	See above.
1989	Yang, Sarkar, and Arvas [39]	See above.
1995	Fu and Dowling [18]	See above.

There are many closely related ideas such as steepest descents or coordinate relaxation that are not discussed here. It is instructive to consider the subtle variations between similar algorithms, as well as the key differences between algorithms that sound the same, but are quite different.

Almost all the algorithmic ideas carry over to the generalized eigenvalue problem, as was recognized by most of the authors. For purposes of exposition, we will only discuss the ordinary eigenvalue problem in this paper. We have chosen to confine our discussion to conjugate gradient methods for the general eigenproblem. The larger history of conjugate gradient methods may be found in the survey by Golub and O'Leary [22]. The search direction choice(s) appear in the annotations.

5 The Differential Geometry Viewpoint for NCG

In a recent paper [14], we proposed an algorithm to perform minimization on the block Rayleigh quotient that has appealing theoretical properties. Our algorithm takes into account both the constraints $Y^T Y = I_p$ and the degeneracy of the function ($R(Y) = R(YQ)$ for orthogonal $p \times p$ matrices Q) in a clean manner. Following the lead proposed by Smith [32, 33], we derived an algorithm based on intrinsic geometry that is practical when expressed in extrinsic coordinates.

Like French-English false cognates (*faux amis*), there are a few terms used both in optimization and differential geometry with somewhat different meanings. This may cause confusion to readers of our paper [14] who are specialists in optimization.

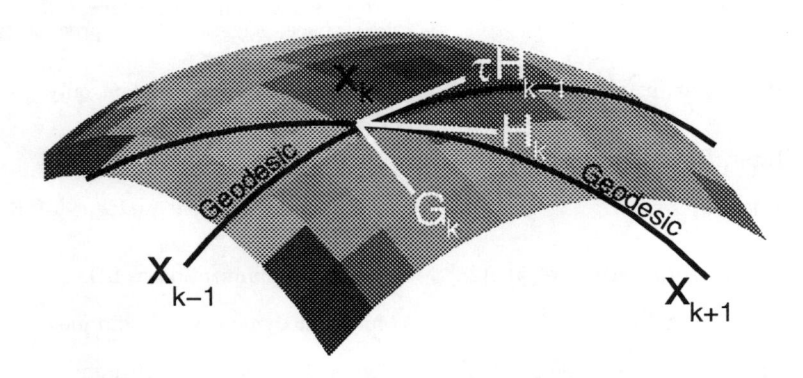

FIG. 1. *Conjugate gradient on Riemannian manifolds.*

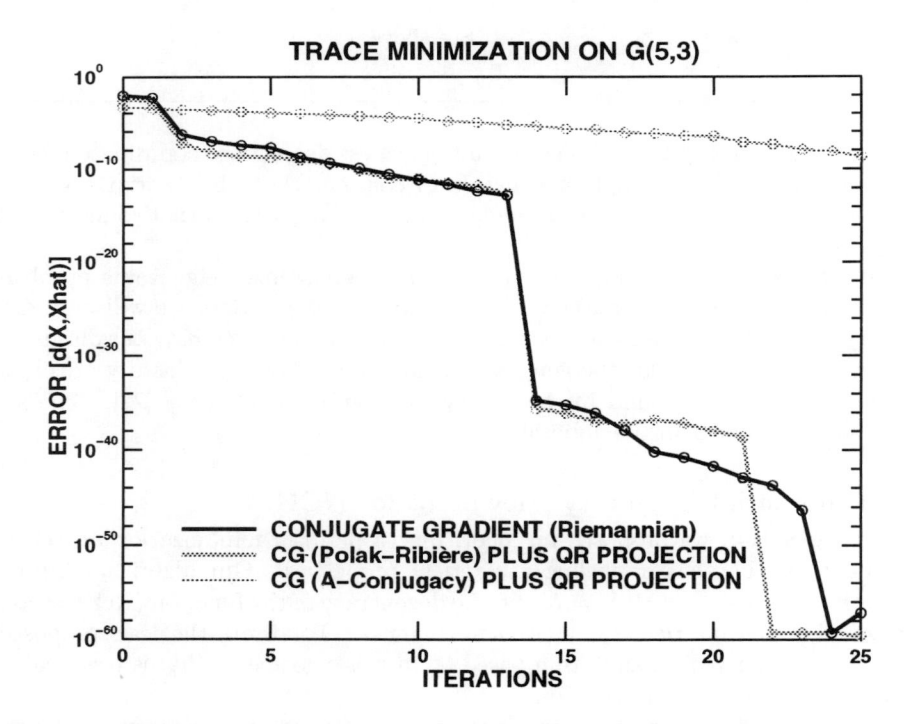

FIG. 2. *Riemannian conjugate gradient, unconstrained block Polak-Ribière, and the A-conjugacy algorithms compared.*

> **Optimization vs. Differential Geometry**
>
> **Metric**
> Optimization: Sometimes the inner product defined by the Hessian or its inverse.
> Diff. Geom.: Any positive definite inner product defined on a tangent space from which all geometrical information about a space may be derived.
>
> **Curvature**
> Optimization: Hessian. Usage refers to non-linearity of the graph of a function with nonzero Hessian.
> Diff. Geom.: A rank four tensor that refers roughly to the non-flatness of a space.

The Grassmann algorithm [14] does not suffer from any of the deficiencies of the other block algorithms for the eigenvalue problem. There is no requirement that the matrix be positive definite, there is no theoretical difficulty related to the singular Hessian, and the algorithm converges quadratically.

The key components of the Grassmann algorithm are the total space/base space point of view, the following of geodesics, and the parallel transportation of tangent vectors. Below we plot a schematic figure that describes conjugate gradient on a Riemannian manifold:

In Figure 2, we show convergence curves in exact arithmetic for three block algorithms. The A-conjugacy algorithm is clearly inferior and not worth considering. The Riemannian algorithm and our own unconstrained version of the block Polak-Ribière algorithm, which approximates the Riemannian CG algorithm up to second order and is less computationally intensive than the full-blown Riemannian algorithm, are quadratically convergent.

6 Beyond Conjugate Gradient

Potential users of a conjugate gradient algorithm may well consider the viewpoint expressed by Jorge Nocedal [26]:

> The recent development of limited memory and discrete Newton methods have narrowed the class of problems for which conjugate gradient methods are recommended. Nevertheless, in my view, conjugate gradient methods are still the best choice for solving very large problems with relatively inexpensive objective functions. They can also be more suitable than limited memory methods on several types of multiprocessor computers.

Though we believe that we have derived the correct NCG algorithm for functions such as the block Rayleigh quotient, it is very possible that the best algorithms for applications in the LDA community and the signal processing community may well be a Newton iteration rather than an algorithm in the conjugate gradient family. Rightly or wrongly, the CG algorithms may yet remain popular for the largest problems because of the simplicity of programming and limited memory requirements.

It is important to understand the relationship between preconditioned conjugate gradient and Newton's method. We sometimes consider the ideal preconditioner to be one that is easily computed, and yet closely resembles the inverse of the Hessian. Multiplying the gradient by the exact inverse of the Hessian is exactly the Newton method. Therefore the Newton method is equivalent to preconditioned conjugate gradient without any use of a previous search direction. One expects that taking advantage of Hessian information, if convenient, would lead to superior convergence.

We now explore Newton's method for invariant subspace computation more closely. Suppose (for simplicity only) that A is symmetric. We would then wish to find matrices Y and B such that

(1) $$AY - YB = 0.$$

The above system is degenerate because there are more unknown variables than equations; constraints of some form or another need to be imposed. One approach is the general

(2) affine constraint $Z^T Y = I$

introduced by Chatelin [5, 6] in which case $B = Z^T A Y$. We observe that the algorithms proposed by Dongarra, Moler, and Wilkinson [13] and Stewart [34]. represent two special choices for the affine constraint matrix Z. In the Dongarra et al. case, Z may be obtained by inverting and transposing an arbitrary $p \times p$ minor of the $n \times p$ matrix Y. In Moler's Matlab notation, Z=zeros(n,p); Z(r,:)=inv(Y(r,:))', where r denotes a p-vector of row indices. For Stewart, $Z = Y(Y^T Y)^{-1}$. Demmel [12, Section3] originally observed that the three algorithms were related. Being linear, the affine constraint allows a direct application of Newton's method without any difficulty.

An alternative choice for the constraint is the

(3) orthogonality constraint $Y^T Y = I$.

Newton's method with this constraint is slightly more complicated because of the nonlinearity. The value of B is now $Y^T A Y$. However the equation $AY - Y(Y^T A Y) = 0$, even with the orthogonality constraints, is degenerate. This makes this problem even more difficult to handle.

The geometric approach to the Grassmann manifold gives an approach to resolving this problem. We do not derive the details here, but the Grassmann point of view on the Newton method starts with the second covariant derivative of the block Rayleigh quotient (see [14]):

(4) $$\mathrm{Hess}(\Delta_1, \Delta_2) = \mathrm{tr}\left(\Delta_1^T A \Delta_2 - (\Delta_1^T \Delta_2) Y^T A Y\right).$$

From there to pick the Newton search direction, we must solve for Δ such that $Y^T \Delta = 0$ in the Sylvester equation

(5) $$\Pi(A\Delta - \Delta(Y^T A Y)) = -G,$$

where $\Pi = (I - YY^T)$ denotes the projection onto the tangent space of the Grassmann manifold, and $G = \Pi A Y$ is the gradient. We then may use this search direction to follow a geodesic.

Equation (5) is really a column by column Rayleigh quotient iteration in disguise. To see this, write $Y^T A Y =: Q \Theta Q^T$, Θ diagonal, and project Equation (5) onto Image(Π). The projected equation is

(6) $$\bar{A}\bar{\Delta} - \bar{\Delta}\Theta = -\bar{G},$$

where the barred quantities are $\bar{A} = \Pi A \Pi$, $\bar{\Delta} = \Pi \Delta Q$ is a column matrix of Ritz vectors [27], and $\bar{G} = GQ$. If $\bar{\Delta}$ is any solution to the above equation, then $\Delta = \bar{\Delta} Q^T$ is a solution to Equation (5). One way to interpret Equation (6) is that we turned the orthogonality constraint $Y^T Y = I$ into an affine constraint $\Delta^T Y = 0$ by differentiating.

Solving for Δ and exponentiating amounts to performing the Rayleigh quotient iteration on each of the Ritz vectors associated with the subspace Y. Therefore, Newton's method applied to the function $\mathrm{tr}Y^TAY$ on the Grassmann manifold converges cubically. This is a generalization of the identification between RQI and Newton's method applied to Rayleigh's quotient on the sphere [32, 33]. This method also has very much in common with Chatelin's method, yet it is the natural algorithm from the differential geometry point of view given the orthogonality constraints $Y^TY = I$.

References

[1] B. M. Alsén, *Multiple step gradient iterative methods for computing eigenvalues of large symmetric matrices*, Tech. Rep. UMINF-15, University of Umeå, 1971.

[2] I. Andersson, *Experiments with the conjugate gradient algorithm for the determination of eigenvalues of symmetric matrices*, Tech. Rep. UMINF-4.71, University of Umeå, 1971.

[3] T. A. Arias, M. C. Payne, and J. D. Joannopoulous, *Ab initio molecular dynamics: analytically continued energy functionals and insights into iterative solutions*, Physical Review Letters, 71 (1992), pp. 1077–1080.

[4] W. W. Bradbury and R. Fletcher, *New iterative methods for solutions of the eigenproblem*, Numerische Mathematik, 9 (1966), pp. 259–267.

[5] F. Chatelin, *Simultaneous Newton's iteration for the eigenproblem*, Computing, Suppl., 5 (1984), pp. 67–74.

[6] ———, *Eigenvalues of Matrices*, John Wiley & Sons, New York, 1993.

[7] H. Chen, T. K. Sarkar, S. A. Dianat, and J. D. Brulé, *Adaptive spectral estimation by the conjugate gradient method*, IEEE Trans. Acoustics, Sppech, and Signal Processing, ASSP-34 (1986), pp. 272–284.

[8] J. K. Cullum, *The simultaneous computation of a few of the algebraically largest and smallest eigenvalues of a large, sparse, symmetric matrix*, BIT, 18 (1978), pp. 265–275.

[9] J. K. Cullum and W. E. Donath, *A block generalization of the symmetric s-step Lanczos algorithm*, Tech. Rep. RC 4845, IBM Research, Yorktown Heights, New York, 1974.

[10] ———, *A block Lanczos algorithm for computing the q algebraically largest eigenvalues and a corresponding eigenspace of large, sparse, real symmetric matrices*, in Proceedings of the 1974 Conference on Decision and Control, Phoenix, Arizona, November 1974, pp. 505–509.

[11] J. K. Cullum and R. A. Willoughby, *Lanczos Algorithms for Large Symmetric Eigenvalue Computations*, vol. 1 Theory, Birkhäuser, Stuttgart, 1985.

[12] J. W. Demmel, *Three methods for refining estimates of invariant subspaces*, Computing, 38 (1987), pp. 43–57.

[13] J. J. Dongarra, C. B. Moler, and J. H. Wilkinson, *Improving the accuracy of computed eigenvalues and eigenvectors*, SIAM J. Num. Analysis, 20 (1983), pp. 46–58.

[14] A. Edelman, T. A. Arias, and S. T. Smith, *Conjugate gradient on the Stiefel and Grassmann manifolds*, submitted to SIAM J. Matrix Anal. Appl., (1995).

[15] R. L. Fox and M. P. Kapoor, *A miminimization method for the solution of the eigenproblem arising in structural dynamics*, in Proceedings of the Second Conference on Matrix Methods in Structural Mechanics, L. Berke, R. M.Bader, W. J. Mykytow, J. S. Przemieniecki, and M. H. Shirk, eds., Wright-Patterson Air Force Base, Ohio, 1969, pp. 271–306.

[16] I. Fried, *Gradient methods for finite element eigenproblems*, AIAA Journal, 7 (1969), pp. 739–741.

[17] ———, *Optimal gradient minimization scheme for finite element eigenproblems*, J. Sound Vib, 20 (1972), pp. 333–342.

[18] Z. Fu and E. M. Dowling, *Conjugate gradient eigenstructure tracking for adaptive spectral estimation*, IEEE Trans. Signal Processing, 43 (1995), pp. 1151–1160.

[19] D. R. Fuhrmann and B. Liu, *An iterative algorithm for locating the minimal eigenvector of a symmetric matrix*, in Proc. IEEE Int. Conf. Acoust., Speech, Signal Processing, 1984, pp. 45.8.1–4.

[20] M. Geradin, *The computational efficiency of a new minimization algorithm for eigenvalue analysis*, J. Sound Vibration, 19 (1971), pp. 319–331.
[21] M. J. Gillan, *Calculation of the vacancy formation energy in aluminium*, Journal of Physics, Condensed Matter, 1 (1989), pp. 689–711.
[22] G. Golub and D. O'Leary, *Some history of the conjugate gradient and Lanczos methods*, SIAM Review, 31 (1989), pp. 50–102.
[23] R. Haimi-Cohen and A. Cohen, *Gradient-type algorithms for partial singular value decomposition*, IEEE Trans. Pattern. Anal. Machine Intell., PAMI-9 (1987), pp. 137–142.
[24] W. Karush, *An iterative method for finding characteristic vectors of a symmetric matrix*, Pacific J. Math., 1 (1951), pp. 233–248.
[25] G. Kresse and J. Hafner, *Ab initio molecular dynamics for liquid metals*, Physical Review B, (1993), pp. 558–561.
[26] J. Nocedal, *Theory of algorithms for unconstrained optimization*, Acta Numerica, (1992), pp. 199–242.
[27] B. N. Parlett, *The Symmetric Eigenvalue Problem*, Prentice-Hall, Inc., Englewood Cliffs, NJ, 1980.
[28] M. C. Payne, M. P. Teter, D. C. Allan, T. A. Arias, and J. D. Joannopoulos, *Iterative minimization techniques for ab initio total-energy calculations: molecular dynamics and conjugate gradients*, Rev. Mod. Phys, 64 (1992), pp. 1045–1097.
[29] A. Perdon and G. Gambolati, *Extreme eigenvalues of large sparse matrices by Rayleigh quotient and modified conjugate gradients*, Comp. Methods Appl. Mech. Engin., 56 (1986), pp. 251–264.
[30] A. Ruhe, *Iterative eigenvalue algorithms for large symmetric matrices*, in Numerische Behandlung von Eigenwertaaufgaben Oberwolfach 1972, Intl, Series Numerical Math. Volume 24, 1974, pp. 97–115.
[31] A. H. Sameh and J. A. Wisniewski, *A trace minimization algorithm for the generalized eigenvalue problem*, SIAM Journal of Numerical Analysis, 19 (1982), pp. 1243–1259.
[32] S. T. Smith, *Geometric Optimization Methods for Adaptive Filtering*, PhD thesis, Harvard University, Cambridge, Massachusetts, 1993.
[33] ———, *Optimization techniques on Riemannian manifolds*, in Hamiltonian and Gradient Flows, Algorithms and Control, A. Bloch, ed., vol. 3 of Fields Institute Communications, Providence, RI, 1994, American Mathematical Society, pp. 113–146.
[34] G. W. Stewart, *Error and perturbation bounds for subspaces associated with certain eigenvalue problems*, SIAM Review, 15 (1973), pp. 752–764.
[35] I. Štich, R. Car, M. Parrinello, and S. Baroni, *Conjugate gradient minimization of the energy functional: A new method for electronic structure calculation*, Phys. Rev. B., 39 (1989), pp. 4997–5004.
[36] M. W. Sung, R. Kawai, and J. Weare, *Packing transitions in nanosized Li clusters*, Physical Review Letter, 73 (1994), pp. 3552–3555.
[37] M. P. Teter, M. C. Payne, and D. C. Allan, *Solution of Schrödinger's equation for large systems*, Phys. Review B, 40 (1989), pp. 12255–12263.
[38] M. A. Townsend and G. E. Johnson, *In favor of conjugate directions: A generalized acceptable-point algorithm for function minimization*, J. Franklin Inst., 306 (1978).
[39] X. Yang, T. P. Sarkar, and E. Arvas, *A survey of conjugate gradient algorithms for solution of extreme eigen-problems of a symmetric matrix*, IEEE Trans. Acoust., Speech, Signal Processing, 37 (1989), pp. 1550–1556.

Chapter 4
Requirements for Optimization in Engineering

D. P. Young,* W. P. Huffman,* R. G. Melvin,*
F. T. Johnson,* C. L. Hilmes,* and M. B. Bieterman*

Abstract

We discuss some of the requirements for the successful application of optimization methodology in aerodynamic design. Current non-optimization design methods are effective, flexible, and inexpensive in practice and there are mathematical and algorithmic difficulties that must be addressed to make optimization based methods competetive and effective. These issues are often not the ones addressed in the numerical optimization literature. We suggest that the generic definition of the problem often used in numerical optimization may be inadequate to achieve the desired practical result.

1 Introduction

Many industrial design processes are fundamentally inverse problems: given some performance characteristic, design a shape or structure that optimizes this characteristic or that achieves this characteristic and perhaps minimizes manufacturing cost. For example in wing design, minimizing weight and drag are usually desirable design goals. However, often the exact and quantifiable definition of "goodness" of the design is difficult to determine at the beginning of the design process.

In aircraft design, direct measures of performance are often achieved by indirect design methods. For example, aerodynamic drag is often addressed by specifying a wing pressure distribution that is known to achieve relatively low drag while retaining certain other desirable features such as a specified span load. The wing geometry is then modified in an attempt to achieve this pressure distribution. This process is not closed in the sense that there may be no wing that achieves a specified pressure distribution while satisfying all geometry, manufacturing, and structural constraints. In addition, the process can be difficult to carry out reliably for complex geometries such as for a wing in the presence of an engine/strut. Nevertheless, this process is a powerful tool in the hands of an experienced engineer. A very significant advantage of this process is that the computational tools used are typically very fast enabling the engineer to rapidly adapt the definition of "goodness" as the design proceeds.

In the last few years, optimization has been shown to have great potential to make these design processes more systematic and (perhaps more importantly) to make them more robust and allow more engineering creativity. Examples of research in this direction can be seen in [14, 15, 17, 23, 22, 26, 8, 9, 25, 1, 6]. The problem is typically formulated as a constrained optimization problem where the aerodynamic and/or structural equations are treated as equality constraints. Using this formulation, classical methods from optimal

---

*The Boeing Company, M/S 7H-96, PO Box 3707, Seattle WA 98124-2207.

control can be used. Often, large numbers of inequality constraints are required (coming for example from manufacturing considerations). Large numbers of design parameters are often necessary. However, the existing algorithms and software in numerical optimization are typically not adequate for this type of problem. Many inverse design problems are ill-posed and require special regularization methods. For example, in compressible fluid dynamics, artificial viscosity must be added to rule out physically impossible expansion shocks. In addition, most aerodynamics and structures codes have discretization and solution methodology that is not readily amenable to optimization. Optimization methods and codes generally treat the analysis code as a "black box" and compute gradients by finite differences. Typically, Hessian information is generated using a recursive update strategy. Methods resulting from this paradigm are typically adequate only for problems with small numbers of design variables.

The difficulties discussed above are particularly troublesome in three space dimensions where iterative methods are usually used in differential equations software. In order to overcome these difficulties so that the full potential of optimization can be realized in aerodynamics and aeroelastics, we have been pursuing a multi-pronged approach. We have been using the TRANAIR aerodynamics code which solves the full potential equation coupled with an integral boundary layer as a testbed to study the algorithmic issues. At the same time, this code provides a prototype capability that can be used in actual engineering design situations where more conventional methods have not been entirely successful. TRANAIR is a natural code for design and optimization because of its use of a Newton solver, solution adaptive grid refinement, and its ability to handle arbitrary geometries. We have recently developed methodology for putting an optimization process "inside" a differential equations solution algorithm that allows the incorporation of solution adaptive grid refinement and the use of "inexact" gradient and Hessian information. This paradigm enables the development of methods that are both more robust and much more efficient for the solution of large optimization problems. It also allows the use of powerful application dependent global damping strategies to aid global convergence such as the use of multiple grids.

The design and optimization capability in TRANAIR has been applied to aerodynamic design on many Boeing airplane projects. Plans are underway to formally include this kind of design methodology in the aerodynamic design process at Boeing and to extend it to more general settings such as aeroelastic design and to the Navier-Stokes equations for fluid flow. However, significant mathematical and technical difficulties remain to be resolved.

2 Issues in Design Optimization

In the aircraft industry, management is placing increasing emphasis on improving the processes used to design and produce aircraft as a way of reducing cost and improving quality [24]. It is no longer adequate to develop a high technology tool and "toss it over the fence" to the user community. In considering design optimization, careful attention must be paid to its place in various engineering design processes.

For example, the current wing design process involves inverse design codes. These codes take a pressure distribution specified using engineering experience that is thought to be desirable and give a geometry that produces this pressure distribution. However, there is no guarantee that the geometry is smooth, that it closes at the trailing edge, or even that it does not cross itself let alone satisfy more sophisticated constraints. Thus any manufacturing and structural constraints must be satisfied using after the fact smoothing of the geometry or modification of the pressure distribution. However, this process in general has worked well having the advantage that it uses small computing resources and allows the engineer to control the process in detail even though it is not totally satisfying from the point of view of numerical optimization.

Compressible computational fluid dynamics (CFD) has fostered a revolution in aircraft design over the last twenty years. At transonic flight conditions where the viscous boundary layer remains attached or contains a moderately separated region, CFD analyses are capable of delivering quantitative answers for the sensitivity due to configuration changes. This CFD capability has led to the development of inverse design methods which have revolutionized the transport airplane design process [13, 7]. However, many difficulties that have not been completely resolved in the analysis case become more evident in the context of design optimization.

Many of these difficulties have been known since the 1950's. For example, Morawetz [19, 20] showed that the problem of transonic potential flow past a fixed profile is not well posed. This is because of the sensitivity of the shock location and supersonic zone to small changes in the boundary or boundary conditions. This problem is somewhat alleviated by the introduction of artificial dissipation which is required to rule out nonphysical expansion shocks. However, this merely makes the problem poorly conditioned and the level of accuracy must be traded against the level of artificial viscosity required to make the problem numerically solvable. It has been known for many years that the full potential equation exhibits non-uniqueness even for simple airfoil shapes giving rise to branching in the lift vs. angle of attack curve for example.

Existence and uniqueness are not as well understood for the steady state Euler and Navier-Stokes equations. However, recently Jameson [18] has demonstrated non-unique solutions to the discrete Euler equations and has discovered branching in the lift vs. angle of attack curve for some airfoils. It is significant that these airfoils were the result of applying numerical optimization.

The Navier-Stokes equations are highly nonlinear and seem to be quite ill-conditioned especially near separation. This can be seen in the special case of integral boundary layer codes such as the one in TRANAIR. It is also well known in supersonic freestream flow that there are whole families of airfoil shapes with the same gross performance characteristics, i.e., lift and drag.

Complicating the difficulties is the fact that there are many situations that appear to give rise to attractive designs that are near or in one of the regions of non-uniqueness or ill-conditioning described above. Since these situations could correspond to unsteady flow in practice or even to large scale oscillations known as flutter (which is unacceptable) it is not clear that the global optimum is what is really desired in aircraft design. It is difficult to formulate the constraints that reliably rule out such situations. These difficulties are illustrated schematically in Figure 1 where u is some control such as angle of attack or surface shape and $I(u)$ is some measure of performance such as drag or lift. The inviscid curve has several local minima as well as a flat area near the left end point. The region of the global minimum on the right represents a shock free airfoil. In this vicinity, there is nonuniqness as well as very high sensitivity to small changes in u. This global minimum is clearly not acceptable from an engineering point of view. The regularized curve is helpful in getting close to the desired minimum.

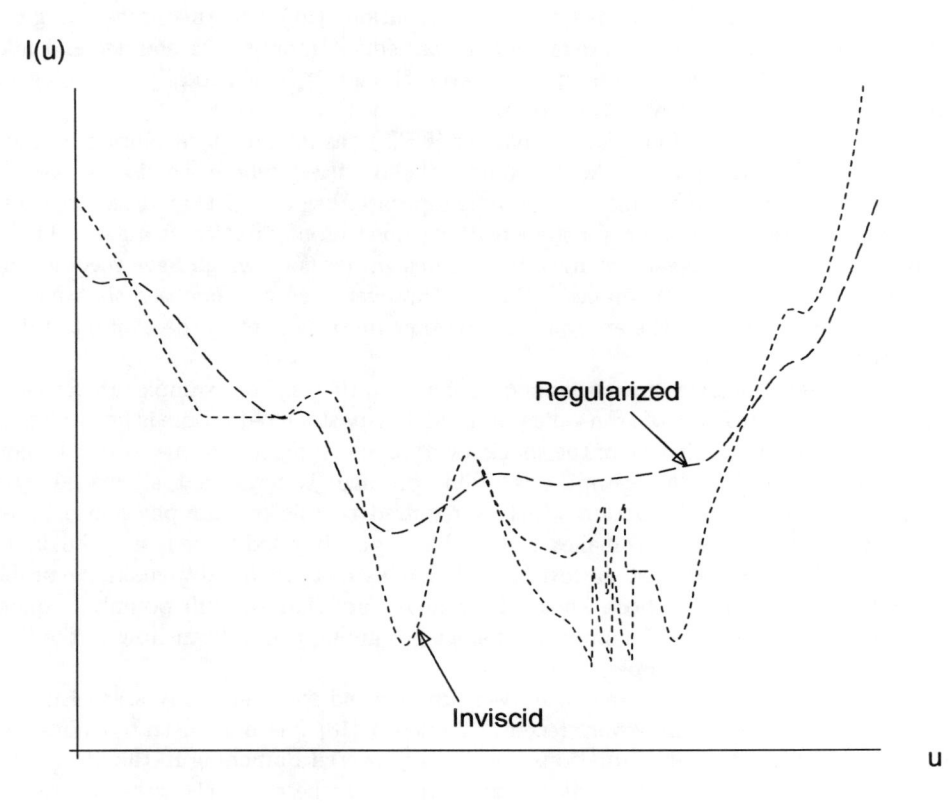

FIG. 1. *Illustrative Schematic of an Aerodynamic Performance Index I as a Function of the Control u.*

3 Newton's Method

Many optimization methods for nonlinear programming are based on Newton's method or inexact Newton's method. There is also a relationship between optimal control methods governed by differential equations and solution of nonlinear systems of equations. The standard optimal control formulations introduce Lagrange multipliers so that the control problem is converted to the problem of solving a nonlinear system of equations. Newton's method is one common method for solving the resulting systems. Thus, it would seem instructive when considering optimization subject to steady state fluid dynamics equations to first consider Newton's method applied to the fluid dynamics equations themselves. Indeed, the difficulties involved in common fluid dynamics models become evident in this context.

The ISES code of Mark Drela [8] as well as Boeing's TRANAIR [30, 16, 4] code use Newton's method to solve inviscid aerodynamic problems with coupled integral boundary layer. Newton's method can be described as follows. Suppose we wish to solve the nonlinear system of equations $F(x) = 0$. Given an initial approximate solution x^0, for $n = 0, 1, 2, \ldots$ until the residual is sufficiently small, set $x^{n+1} = x^n + \lambda(\delta x^{n+1})$ where δx^{n+1} is the solution of the linear system

(1) $$\overline{F}_{x^n}(\delta x^{n+1}) = -F(x^n)$$

and λ is a step length to be determined. Here $\overline{F}_{x^n}$ is the Jacobian for F linearized about x^n. The step length λ is selected in various ways. One simple strategy is to use a line search, i.e., to minimize $\|F(x^{n+1})\|$ in some appropriate norm.

When iterative methods are used, the system (1) is often solved only approximately, i.e., δx^{n+1} satisfies $\|\overline{F}_{x^n}(\delta x^{n+1})+F(x^n)\| < \eta\|F(x^n)\|$. This is called an inexact Newton method and can be shown to converge linearly if η is held constant. However, in our experience, the total CPU time for an inexact Newton method is usually a factor of 2 to 4 less than that for a quadratically convergent Newton method. For large systems iterative methods usually require a preconditioner N in which case the system $N^{-1}\overline{F}_{x^n}(\delta x^{n+1}) = -N^{-1}F(x^n)$ is solved.

Newton's method is rarely globally convergent. In the full potential case, an easily computed initial iterate is freestream flow. With this initial guess, Newton's method works well only for small problems or those without shocks. Damping of Newton's method must be introduced for large transonic problems to prevent stagnation of convergence. This can be accomplished in several ways. TRANAIR uses a line search strategy based on the golden section. This strategy was chosen because of its simplicity and ease of implementation. More sophisticated strategies such as trust region methods [10] have been tried in the past without success.

A more powerful damping strategy is called viscosity damping in which a dissipation parameter is introduced. Initially, the problem is modified by increasing the amount of artificial viscosity in regions of supersonic flow and applying it to a larger part of the flow field. The continuation parameter is reduced in discrete steps, two to four usually being sufficient, until the desired level of dissipation is reached. This continuation process has the effect of locating supersonic zones and shock positions fairly early in the process, even though the shocks are quite smeared.

An even more effective damping strategy is grid sequencing. Grid sequencing consists of solving the problem first on a coarse grid, interpolating this solution to a finer grid, and using the interpolant as the initial guess to solve the finer grid problem. Such a sequence of grids is generated automatically in a solution adaptive grid method. More details on these strategies can be found in [3, 29].

The effect of these various global strategies can be seen in the case of the ONERA M6 wing at $M_\infty = 0.84$ and $\alpha = 3.06°$. Figure 2 shows convergence histories with a freestream initial guess, viscosity damping with freestream initial guess, and grid sequencing for various upwinding methods for this case. Each linearized problem generated by Newton's method was solved to at least 2 digits in the unpreconditioned residual and preconditioners were recomputed whenever convergence of the linearized problems slowed. A very significant increase in cost for the slowly converging cases is due to this need to frequently recompute the preconditioner. The Mach profiles shown are at approximately 70% of semi-span. We note that there does seem to be a numerical limit to how refined the grid can be in a strong shock wave without causing Newton's method to stagnate; even when the damping strategies discussed here are used. Fortunately, this limit is well beyond what is needed for engineering calculations and is probably beyond physical realizability. However, this does illustrate the fact that as truncation error goes to zero, the problem becomes more and more poorly conditioned.

In two space dimensions or in cases of boundary layer coupling or in more difficult

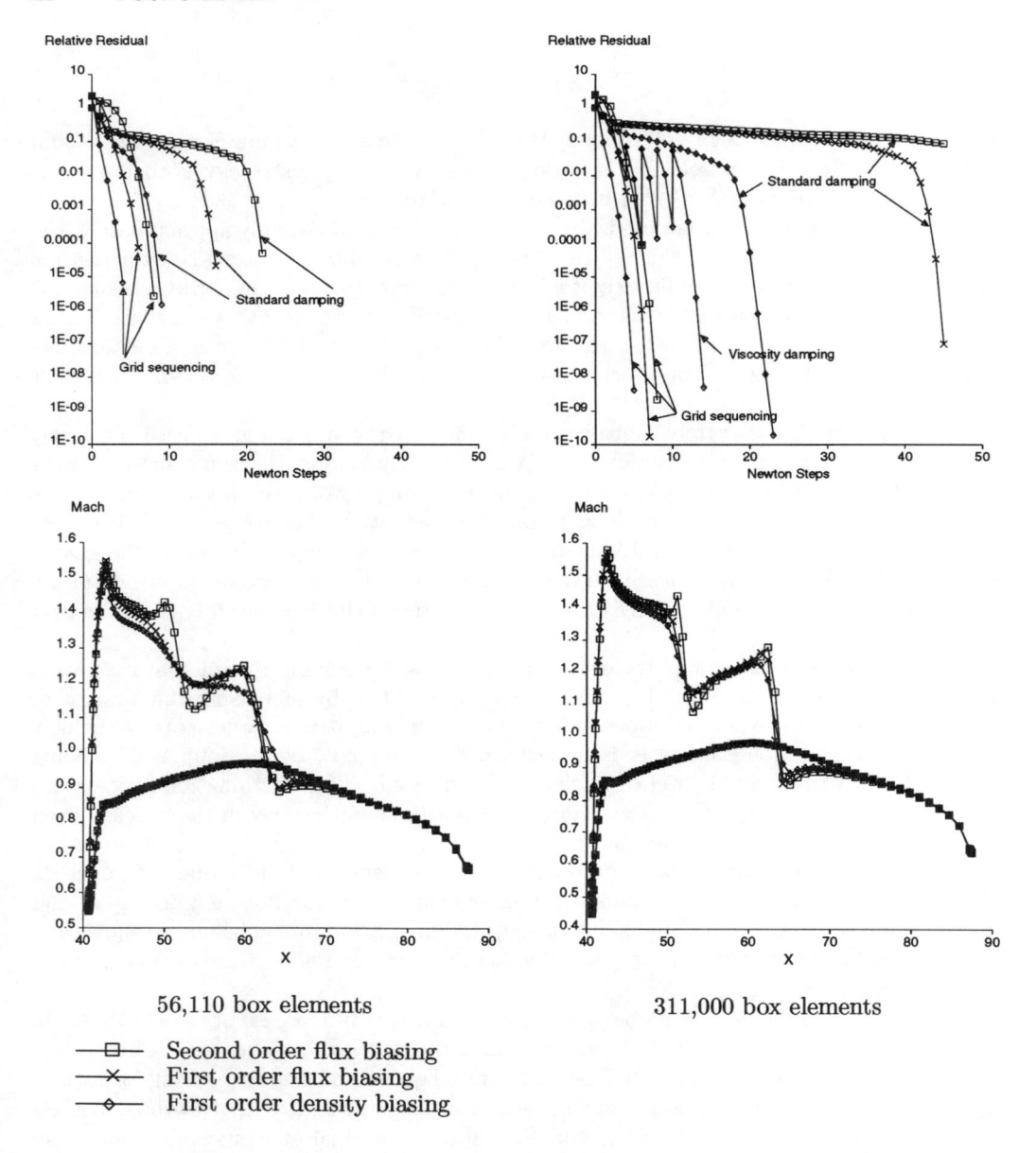

FIG. 2. *Convergence of Newton's Method for ONERA M6 Wing, $M_\infty = 0.84, \alpha = 3.06°$. Freestream Initial Guess vs. Grid Sequencing vs. Viscosity Damping.*

cases involving strong shocks the differences between various global strategies documented above become even more pronounced. As discussed in [28] these lessons about global damping strategies seem to carry over to aerodynamic optimization and to suggest that difficult analysis problems may prove a useful testbed for determining the usefulness of global optimization and/or multi-disciplinary coupling strategies. All of the more effective damping strategies outlined above can be applied in optimization situations using the optimization formulation discussed in [28].

4 Design and Optimization Formulation

By placing an optimization method "inside" a solution adaptive method for solving PDE's, one can allow the use of solution adaptive grid refinement and inexact gradient and Hessian information. This makes possible the efficient solution of design problems while removing many of the restrictions thought to exist for numerical optimization methods applied to aerodynamic or aeroelastic optimization. On each grid of a solution adaptive grid run, a simplified and regularized optimization sub-problem is defined and solved using a constrained optimization method. In the case of a least squares objective, the LSSOL package is used [11]. In other cases, the NPSOL nonlinear programming package is used [12]. On each grid, the method used is a standard reduced gradient method [10]. An important feature of the method is the use of a transpiration boundary condition [16] on a fixed surface to approximate the effect of actually moving the designed surface. Complete details of the algorithm can be found in [28].

The constrained design problem can be stated as the problem of minimizing a scalar objective function $I(X, u)$ subject to the constraint that $F(X, u) = 0$. In this formulation, u are the design variables and X are the state variables, i.e., variables that enter the problem through the constraint equations and $F(X, u)$ are the differential equations governing the problem. The necessary conditions for optimality are often formulated by introducing the Lagrange multipliers λ as independent variables [5] and introducing the Lagrangian function. As discussed by numerous authors [5], in the cases where $\partial F/\partial X$ is invertible for reasonable values of u, the Lagrange multipliers can be eliminated from the formulation by noting that

$$\lambda = -\left(\left(\frac{\partial F}{\partial X}\right)^*\right)^{-1}\left(\frac{\partial I}{\partial X}\right)^*. \tag{2}$$

The necessary conditions for optimality can be stated as:

$$\begin{align}
\frac{dI}{du} &\equiv -\left(\frac{\partial I}{\partial X}\right)\left(\frac{\partial F}{\partial X}\right)^{-1}\left(\frac{\partial F}{\partial u}\right) + \frac{\partial I}{\partial u} \\
&= \lambda^*\left(\frac{\partial F}{\partial u}\right) + \frac{\partial I}{\partial u} = 0 \\
F(X, u) &= 0.
\end{align} \tag{3}$$

The quantity dI/du defined above is often called the reduced gradient. The necessary conditions can now be formulated either in terms of solving an adjoint problem (2) for λ or directly by solving the linearized state equations using multiple right hand sides to compute each column of the n by m sensitivity matrix

$$\frac{\partial X_i}{\partial u_j} = \left(\frac{\partial F_k}{\partial X_i}\right)^{-1}\left(\frac{\partial F_k}{\partial u_j}\right) \tag{4}$$

The application of an inexact Newton method to solve this problem as well as the role of inexactness and solution adaptive grid refinement is discussed in [28]. For purposes of this discussion, it is sufficient to observe that to solve the optimization subproblem on each grid it is necessary to linearize any constraints, $c(X, u)$, involving the state variables. Assuming that the sensitivities, $\partial X_i/\partial u_j$ are known, this is done using the reduced gradient formula given in the first part of equation (3):

$$\text{(5)} \qquad \frac{dc}{du_i} = -\sum_{k=1}^{n} \frac{\partial c}{\partial X_k} \frac{\partial X_k}{\partial u_i} + \frac{\partial c}{\partial u_i}$$

These linearizations could also be computed using one adjoint solve for each constraint as described above for the objective function. In any case, when constraints on the state variables are present such as if one wished to specify the aerodynamic span loading for a wing, the problem of solving linear systems with the same left hand side but multiple right hand sides is encountered. This requirement is not unique to our formulation but holds for any reduced gradient method based on equation (3).

5 Global Convergence In Aerodynamic Design

In this section, we discuss the effect of applying the damping strategies discussed for Newton's method above to the case of aerodynamic optimization. In [28], the insensitivity of global convergence to reasonable levels of "inexactness" was demonstrated. In this section, we use a least squares test case from that paper.

The objective function is

$$I = \frac{1}{2} \sum_{k=1}^{q} W_k [(c_p)_k - (\hat{c}_p)_k]^2$$

where $(\hat{c}_p)_k$ is a target pressure obtained from analyzing an airfoil that is a slightly modified section from the ONERA M6 wing at freestream Mach number $M_\infty = 0.75$ and angle of attack $\alpha = 2.0°$. The index k runs over the surface panel corner points. The baseline airfoil is a Joukowski airfoil at the same freestream Mach number and zero degrees angle of attack resulting in a symmetric subsonic flow field. The geometric mode shapes used to perturb the baseline geometry are Tchebychev polynomials as described in [16] and the weights are the local surface panel lengths. In this case not only the thickness and camber of the airfoil change but also the angle of attack. Figure 3 shows that the transpiration design predicted surface pressure with 30 terms is quite close to the target. As would be expected in such a case where the leading edge is moved a considerable distance, there is some error in the transpiration model. This can be seen by comparing the designed geometry and the target in Figure 3. The transpiration approximation and constraints used are described in detail in [16].

For all the runs discussed below, 30 Tchebychev polynomial terms were used as design mode shapes changing the shape of the airfoil. Solution adaptivity results in a highly locally refined grid which is frozen after grid 11 in order to study asymptotic convergence. Figure 3 shows grids 3 and 11, respectively, for the baseline run. Because of the use of solution adaptivity, the grids generated are not identical for the cases described here; even though they are close. For the first case (the standard way the code is run), the first 11 grids had 173, 335, 662, 1167, 2218, 4219, 5577, 6405, 7398, 8515, 9264 finite elements respectively. The cycle of nonlinear flow solve and optimization problem setup and solution is repeated on this final grid until convergence is achieved. In a normal solution adaptive run, fewer grids are used and the rate at which the numbers of elements grow from grid to grid is greater. The design process started on the third grid. The first run is a standard least squares optimization run. For the second run, the sensitivities were calculated by linearizing a

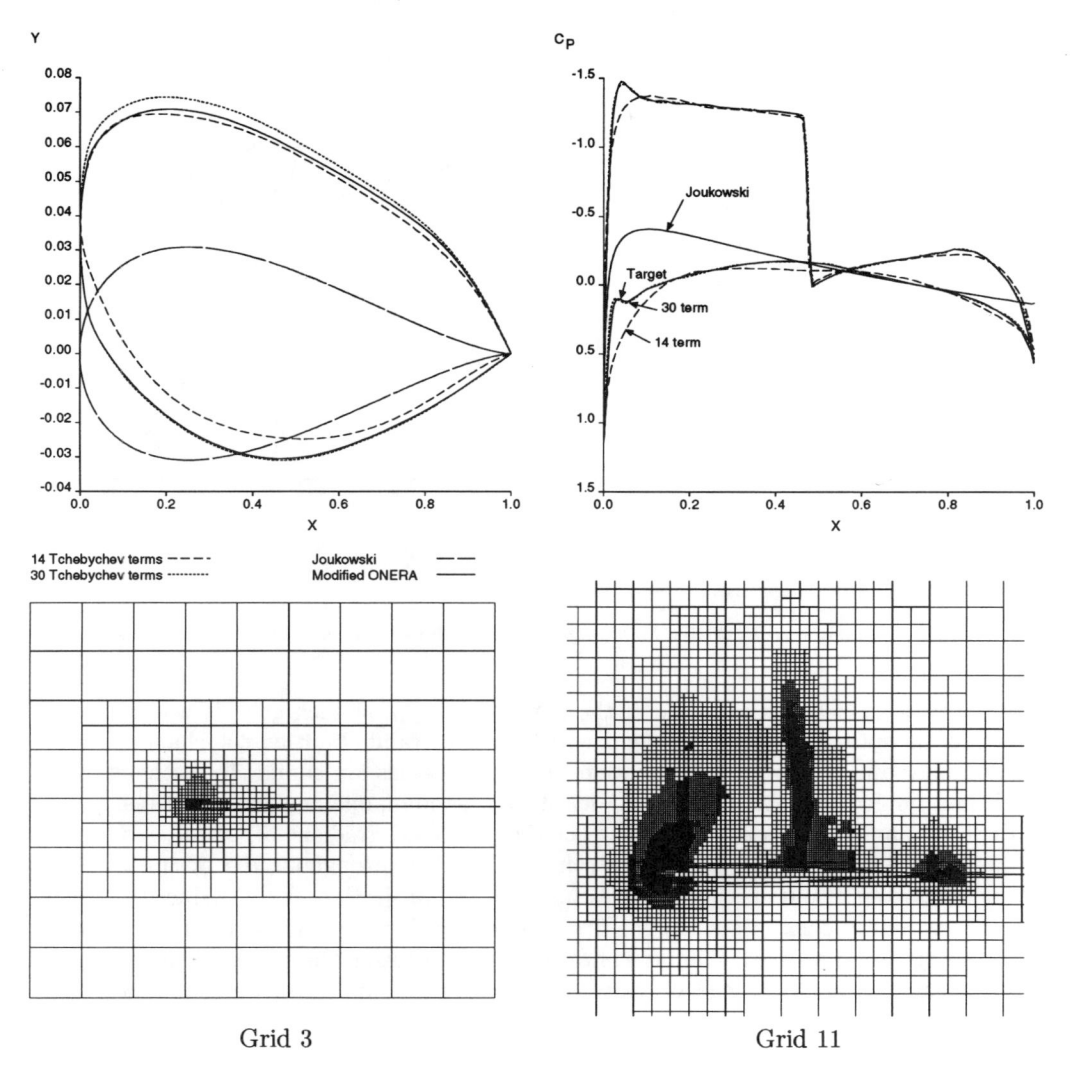

FIG. 3. *Geometries, Pressure Distributions, and Grids for Joukowski to Modified ONERA M6 Airfoil Design.*

flow problem in which additional artificial viscosity was added and the cutoff Mach number lowered as discussed in [29]. For the third run, "good" values of the design variables were inserted into a standard run on grid 7. A fourth run inserted the initial geometry on the final grid of an optimization run. This grid was then repeated a large number of times. This run simulates what would happen without the benefit of grid sequencing and its resulting effect on global convergence. Figure 4 shows the objective function values as a function of grid number. The reason for the jumps is that the value is shown twice for each grid, once before the least squares step and once afterward. This second value is a linearized estimate only. Thus, jumps in objective value from its estimate on the preceding grid can indicate either strong nonlinear effects or significant changes in the solution adaptive grids.

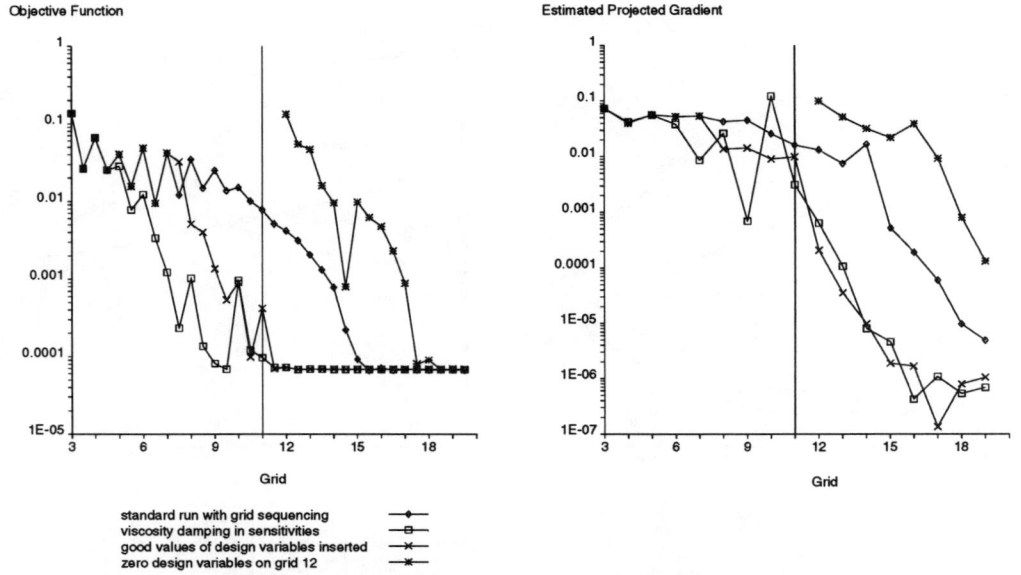

FIG. 4. *Convergence Histories for Joukowski to Modified ONERA M6 Airfoil Design.*

In all cases, to achieve convergence of the flow solver step on each grid, it was necessary to modify the optimization subproblems with flow constraints limiting the change in local Mach number and change in transpiration. These strategies are similar to methods found useful (and implemented in TRANAIR) for steady aerodynamic analysis problems.

Figure 4 shows values of an estimate of the projected gradient of the objective, i.e., $\delta u \cdot (dI/du) / \| \delta u \|$. If the same constraints are active throughout the optimization or least squares step of our method, this estimate is the projection of the reduced gradient onto the inequality constraints. In the first design run discussed above (and in the other three runs discussed as well), there is some difficulty in locating the shock position on early grids. Using additional artificial viscosity totally cures this difficulty; which is much more severe without grid sequencing as shown in the figures.

6 Quadratic Programming Approach

Unfortunately, the excellent convergence results reported above for least squares problems do not seem to carry over to scalar objective functions as discussed and documented in detail in [27]. The difficulties were shown to **NOT** be related to the accuracy with which sensitivities were computed or to the presence of numerical "noise" in the objective function or constraint subroutines. It is believed that some of the difficulties are due to the presence of zero curvature at the solution and directions of negative curvature away from the solution. We have also documented that deficiencies in existing optimization software may be partly responsible [27, 2].

In the absence of Hessian information, the BFGS Hessian update algorithm implemented in NPSOL is equivalent to unpreconditioned conjugate gradient for a quadratic program [21]. It is well known that this is a slowly converging algorithm for problems of moderate dimension and/or conditioning. Inequality constraints seem to further aggravate the difficulty. We have observed these limitations of NPSOL in our experiments with scalar objective functions [27, 2]. Shown in Table 1 are CPU times for application of NPSOL to various drag minimization subproblems generated by TRANAIR. All results are on the final grid of an optimization run and were cold starts in the sense that the initial Hessian was the identity matrix. However, grid sequencing was used in all runs with solution adaptive gridding to provide good initial values for the design variables. We note that these subproblems are totally consistent to machine precision and that in both the 3D cases NPSOL terminated without reporting convergence even though the result seemed upon inspection to be "moving in the right direction". The 3D problems arise from drag minimization for a high speed civil transport (HSCT) configuration. The resulting function box for the subproblem is very inexpensive in CPU time since the linearized surface velocity sensitivities are precomputed and stored. Thus a function evaluation for the subproblem is about 1000 to 10000 times less expensive than a nonlinear flow solve. Thus, the subproblem would seem to be ideal for such numerical optimization techniques. From the table, it appears that the solution time of NPSOL is growing at an unacceptable rate as the problem size increases. This is clear when one considers that aerodynamicists would like to be able to solve such problems with 500 to 1000 design variables.

For the subproblems considered here, the design mode shapes used to perturb the aerodynamic surfaces were defined by standard quintic spline fits to predefined point movements. The point movements were the design variables. In some cases, slopes at endpoints were also used as design variables for the spline fits. We have subsequently verified that using a mode shape approach in which the movement of the geometry is a linear combination of design variables multiplied by predetermined mode shapes results in better conditioning of the resulting optimization subproblems. Thus, it would seem that the choice of design variables can dramatically impact conditioning as well; even for aerodynamic problems constrained away from regions of non-uniqueness. It further appears that simple linear combinations of mode shapes such as those defined by Hicks lead to superior performance of the overall method even though some wiggles or irregularities may be introduced in the designed geometric shapes.

TABLE 1
Performance of NPSOL for drag minimization.

Case	2D airfoil	2D airfoil	3D HSCT	3D HSCT
Grid Size (Number of State Variables)	10,700	10,260	384,800	384,400
# of Design Variables	8	16	40	80
# of Geometry Constraints	192 thick	192 thick	3200 curv	3200 curv
# of Flow Constraints	192	192	3200	3200
Total CPU for Run	628	322	6700	10811
CPU for Sensitivities			900	1618
CPU for NPSOL	12	64	1623	7410
# of Major NPSOL Iterations	9	8	40	81

It is also well known that conjugate gradient, when considered as a method for solving systems of linear equations, has stringent consistency requirements. Each computation of the residual (gradient for a quadratic program) must be completely consistent and the step must be determined in a way that is consistent with these residual (or gradient) calculations. Translating this requirement: if an optimization algorithm is based on the objective function (as are most general optimization codes) then there must be consistency between the gradient evaluations and the objective function values used in the line search. In the case of an approximate Hessian, the quadratic objective function itself is defined in terms of the Hessian and gradient, assuring the consistency required for the attractive convergence properties of methods such as conjugate gradient. As discussed in [27, 2], significant improvements in performance are expected to result from the use of an approximate Hessian calculation in a direct quadratic programming approach.

A further advantage of such a method is immediate information about Hessian conditioning, zero eigenvalues, etc. This information we expect to be invaluable in engineering situations where the well-posedness of the problem is in question. It will then be possible to determine how to modify the mathematical problem to be more amenable to numerical solution.

References

[1] O. Baysal and M. E. Eleshaky, *Preconditioned Domain Decomposition Scheme for Three-Dimensional Aerodynamic Sensitivity Analysis*, AIAA Journal, 32 (1994), pp. 2489–2491.

[2] J. T. Betts, W. P. Huffman, and D. P. Young, *An Investigation of Algorithm Performance for Aerodynamic Design Optimization*, Technical Report BCSTECH-94-061, Boeing Computer Services, Dec. 1994.

[3] M. B. Bieterman, J. E. Bussoletti, C. L. Hilmes, F. T. Johnson, R. G. Melvin, and D. P. Young, *An Adaptive Grid Method for Analysis of 3D Aircraft Configurations*, Computer Methods in Applied Mechanics and Engineering, 101 (1992), pp. 225–249.

[4] M. B. Bieterman, R. G. Melvin, F. T. Johnson, J. E. Bussoletti, D. P. Young, W. P. Huffman, C. L. Hilmes, and M. Drela, *Boundary Layer Coupling in a General Configuration Full Potential Code*, Technical Report BCSTECH-94-032, Boeing Computer Services, Aug. 1994.

[5] A. E. Bryson and Y. C. Ho, *Applied Optimal Control: Optimization, Estimation, and Control*, Hemisphere, New York, 1975.

[6] G. W. Burgreen and O. Baysal, *Three-Dimensional Aerodynamic Shape Optimization of Supersonic Delta Wings*. AIAA Paper 94-4271-CP, Sept. 1994.

[7] R. L. Campbell and L. A. Smith, *A Hybrid Algorithm for Transonic Airfoil and Wing Design*. AIAA Paper 87-2552, 1987.

[8] M. Drela, *Viscous and Inviscid Inverse Schemes Using Newton's Method*, in Special Course on Inverse Methods for Airfoil Design for Aeronautical and Turbomachinery Applications, AGARD Report No. 780, 1990.

[9] ———, *Design and Optimization Method for Multi-element Airfoils*. AIAA Paper 93-0969, 1993.

[10] R. Fletcher, *Practical Methods of Optimization*, John Wiley and Sons, New York, second ed., 1987.

[11] P. E. Gill, S. J. Hammerling, W. Murray, M. A. Saunders, and M. A. Wright, *User's Guide for LSSOL (Version 1.0): A FORTRAN Package for Constrained Linear Least-Squares and Convex Quadratic Programming*, Stanford University Technical Report, Department of Operations Research, 1986.

[12] P. E. Gill, W. Murray, M. A. Saunders, and M. A. Wright, *User's Guide for NPSOL (Version 4.0): A FORTRAN Package Nonlinear Programming*, Stanford University Technical Report SOL86-2, Department of Operations Research, 1986.

[13] M. I. Goldhammer and F. W. Steinle, *Design and Validation of Advanced Transonic Wings Using CFD and Very High Reynolds Number Wind Tunnel Testing*, in ICAS Proceedings, 1990, pp. 1028–1042.

[14] R. M. Hicks and P. A. Henne, *Wing Design by Numerical Optimization*. AIAA Paper 79-0008, 1979.

[15] R. M. Hicks, E. M. Murman, and G. N. Vanderplaats, *An Assessment of Airfoil Design by Numerical Optimization*, Tech. Rep. NASA TM X-3092, NASA Ames Research Center, July 1974.

[16] W. P. Huffman, R. G. Melvin, D. P. Young, F. T. Johnson, J. E. Bussoletti, M. B. Bieterman, and C. L. Hilmes, *Practical Design and Optimization in Computational Fluid Dynamics*. AIAA Paper 93-3111, July 1993.

[17] A. Jameson, *Aerodynamic Design via Control Theory*, Journal of Scientific Computing, 3 (1988), pp. 233–260.

[18] ———, *Airfoils Admitting Non-unique Solutions of the Euler Equations*. AIAA Paper 91-1625, June 1991.

[19] C. S. Morawetz, Comm. Pure Appl. Math., 10 (1957), p. 107.

[20] ———, Comm. Pure Appl. Math., 11 (1958), p. 129.

[21] L. Nazareth, *A Relationship Between the BFGS and Conjugate-Gradient Algorithms and its Implications for New Algorithms*, SIAM Journal on Numerical Analysis, 16 (1979), pp. 794–800.

[22] O. Pironneau, *Optimal Shape Design for Aerodynamics*, in Optimum Design Methods for Aerodynamics, AGARD Report No. 803, Nov. 1994, pp. 6.1–6.40.

[23] J. Reuther and A. Jameson, *Aerodynamic Shape Optimization of Wing and Wing-body Configurations Using Control Theory*. AIAA Paper 95-0123, Jan. 1995.

[24] P. E. Rubbert, *CFD and the Changing World of Airplane Design*. AIAA Wright Brothers Lecture, Sept. 1994.

[25] L. L. Sherman, A. C. Taylor III, L. L. Green, P. A. Newman, G. W. Hou, and V. M. Korivi, *First- and Second-Order Aerodynamic Sensitivity Derivatives via Automatic Differentiation with Incremental Iterative Methods*. AIAA Paper 94-4262 CP, Sept. 1994.

[26] S. Ta'asan, G. Kuruvila, and M. D. Salas, *Aerodynamic Design and Optimization in One Shot*. AIAA Paper 92-005, Jan. 1992.

[27] D. P. Young, W. P. Huffman, M. B. Bieterman, R. G. Melvin, F. T. Johnson, C. L. Hilmes, and A. R. Dusto, *Issues in Design Optimization Methodology*, Technical Report BCSTECH-94-007, Boeing Computer Services, Jan. 1994.

[28] D. P. Young, W. P. Huffman, R. G. Melvin, M. B. Bieterman, C. L. Hilmes, and F. T. Johnson, *Inexactness and Global Convergence in Design Optimization*. AIAA Paper 94-4386, Sept. 1994.

[29] D. P. Young, R. G. Melvin, M. B. Bieterman, F. T. Johnson, and S. S. Samant, *Global Convergence of Inexact Newton Methods for Transonic Flow*, International Journal for Numerical Methods in Fluids, 11 (1990), pp. 1075–1095.

[30] D. P. Young, R. G. Melvin, M. B. Bieterman, F. T. Johnson, S. S. Samant, and J. E. Bussoletti, *A Locally Refined Rectangular Grid Finite Element Method: Application to Computational Fluid Dynamics and Computational Physics*, Journal of Computational Physics, 92 (1991), pp. 1–66.

Chapter 5
On Iterated-Subspace Minimization Methods for Nonlinear Optimization[*]

A. R. Conn[†], Nick Gould[‡], A. Sartenaer[§] and Ph. L. Toint[§]

Abstract

We consider a class of Iterated-Subspace Minimization (ISM) methods for solving large-scale unconstrained minimization problems. At each major iteration of such a method, a low-dimensional manifold, the iterated subspace, is constructed and an approximate minimizer of the objective function in this manifold then determined. The iterated subspace is chosen to contain vectors which ensure global convergence of the overall scheme and may also contain vectors which encourage fast asymptotic convergence. We demonstrate that this approach can sometimes be very advantageous and indicate the general performance on a collection of large problems. Moreover, comparisons with a limited memory approach and LANCELOT are made.

Keywords: Unconstrained optimization, large-scale computation

Mathematics Subject Classifications: 65K05, 90C30

1 Introduction

In this paper, we consider finding a local solution of the unconstrained minimization problem,

(1.1) $$\underset{x \in \Re^n}{\text{minimize}} \ f(x),$$

where we assume, for simplicity, that the objective function $f \in C^2$. We are particularly interested in the case where n is sufficiently large that methods appropriate for small problems — such as those which might maintain a dense factorization of a suitable approximation of the Hessian matrix, see, for example, Gill et al. (1981), Dennis and Schnabel (1983) and Fletcher (1987) — are impractical. We are especially interested

---

[*]This work was supported by the Belgian National Fund for Scientific Research. The research of Conn and Toint was supported in part by the Advanced Research Projects Agency of the Department of Defense and was monitored by the Air Force Office of Scientific Research under Contract No F49620-91-C-0079. The United States Government is authorized to reproduce and distribute reprints for governmental purposes notwithstanding any copyright notation hereon.

[†]IBM T.J. Watson Research Center, P.O.Box 218, Yorktown Heights, NY 10598, USA. Email: arconn@watson.ibm.com.

[‡]Central Computing Department, Rutherford Appleton Laboratory, Chilton, Oxfordshire, OX11 0QX, England. Email: n.gould@letterbox.rl.ac.uk. Current reports available by anonymous ftp from the directory "pub/reports" on joyous-gard.cc.rl.ac.uk (internet 130.246.9.91).

[§]Department of Mathematics, Facultés Universitaires ND de la Paix, 61 rue de Bruxelles, B-5000 Namur, Belgium, EC. Email: as@math.fundp.ac.be or pht@math.fundp.ac.be. Current reports available by anonymous ftp from the directory "pub/reports" on thales.math.fundp.ac.be (internet 138.48.4.14).

in solving large problems for which exact second derivatives are explicitly available. At variance with a popular misconception, such problems frequently arise in practice, as is evident when one considers, amongst others, the large CUTE collection (see Bongartz et al., 1995).

We are primarily concerned with the commonly occurring case in which the cost of evaluating the value of the objective function and its derivatives, at a given point x, is less significant than the cost of solving, for instance, the Newton equations. Our experience with the large-scale nonlinear optimization package **LANCELOT** (see Conn et al., 1992) has been that it is the linear-algebra cost which tends to dominate when solving a significant number of widely differing application problems (see, Conn et al., 1993 and Conn et al., 1996). Thus, it would appear desirable in these cases to attempt to reduce the linear-algebra costs, even if this results in an increase in the number of objective function evaluations.

The most common methods for unconstrained minimization either determine a search direction followed by a linesearch or use the trust-region approach (see, for example, Dennis and Schnabel, 1983). In the former case, a simple model of the underlying objective function is constructed in order to determine the search direction. By contrast, in the latter case, an approximate minimizer of the model within a restricted domain (the trust region) is determined. This model minimizer is then used as a prediction of the actual minimizer of the true objective. In a trust-region method, success of this process is measured by comparing the model and true function values at the predicted minimizer. In linesearch methods, the true function is used to establish a step size. Thus, both of these approaches may be considered to perform their multi-dimensional work with respect to a model whilst probing the true function uni-dimensionally. Of course, the model does make use of the true function and perhaps its derivatives — maybe at more than a single point.

In this paper, we take the view that the above schemes are quite wasteful, given the amount of information that may have been accrued during the (approximate) minimization of the model. In particular, the model may have been sampled in a number of potentially interesting directions, of which only the aggregate direction is normally considered to be of significance.

We also believe that, provided function and derivative values are inexpensive to compute relative to the linear-algebra costs, an (approximate) low-dimensional minimization is a relatively simple calculation. Indeed, we feel that there is high-quality, robust, general-purpose software readily available for the small-scale unconstrained minimization problem, and that such software is normally capable of solving problems of modest dimensions - say up to a hundred variable problems - extremely fast on current workstations *provided that* function evaluation is cheap. Of course there are, and will continue to be, small-scale problems which are challenging, because they are so nonlinear that algorithms implemented in fixed, finite precision arithmetic are unsuccessful, but in our experience such examples occur rarely in practice.

Thus, in this paper, we propose methods which aim to investigate the *true* objective function in a space larger than the one-dimensional space which is normally associated with linesearch or trust-region methods. We do this knowing that, so long as the space is relatively modest, the approximate multi-dimensional minimization will still be a manageable calculation. Moreover, by carefully choosing the space that we investigate, we hope to reduce significantly the linear-algebra costs while still maintaining global, and fast asymptotic, convergence.

Previous work along the lines we are considering includes Cragg and Levy (1969) who propose minimizing in a k-dimensional subspace, which includes the steepest descent direction. The remaining k-1 directions are the steps from the k-1 previous outer iterations. See also, for example, Dennis and Turner (1987), Dixon *et al.* (1984), Miele and Cantrell (1969), and Vinsome (1976). The first minimizes a convex quadratic on a subspace by adding, at each iteration, an extra vector that is not dependent on the existing subspace. They thus provide a uniform framework for a wide class of conjugate directions. The second minimizes the objective function over a grid of 4096 points, generated from a four-dimensional subspace. This subspace is defined using the steepest descent direction, the Newton direction, and two others directions which are combinations of the two previous steps.

A particular form of these ideas has been given by Saad (1990) for the solution of nonlinear systems of equations. Here, a sequence of iterates are generated as least-squares solutions to the equations in suitable Krylov subspaces. The principal difference is that, in Saad's proposal, the entire Krylov subspace generated is used, while, as we shall see, this is in general quite unnecessary. Another example, that also is in the context of nonlinear equations is that of Kaporin and Axelsson (1995). Although developed independently, many of the ideas are quite similar to those of this paper, but the emphasis is rather different.

Given an initial estimate of the solution to (1.1), $x^{(0)}$, and an iteration count, k, set initially to zero, a prototype algorithm for the proposed method might be as follows:

1. Stop with the solution estimate $x^{(k)}$ if convergence tests are satisfied.

2. Determine a full-rank subspace matrix $S^{(k)} \in \Re^{n \times s^{(k)}}$, where $s^{(k)} \ll n$.

3. Approximately solve the $s^{(k)}$-dimensional minimization problem

$$(1.2) \qquad \minimize_{y \in \Re^{s^{(k)}}} f(x^{(k)} + S^{(k)} y),$$

set

$$(1.3) \qquad x^{(k+1)} = \text{(approximate)} \arg\min_{y \in \Re^{s^{(k)}}} f(x^{(k)} + S^{(k)} y),$$

replace k by $k+1$ and return to step 1.

We refer to such a method as *Iterated-Subspace Minimization* or ISM for short. This is, of course, a multi-dimensional subspace analog of the unidimensional-subspace linesearch method.

We are interested in the following issues:

- What is a good choice for $s^{(k)}$?

- How do we determine the *Iterated-Subspace* matrix $S^{(k)}$?

- What do we mean by "approximate" in the problem (1.3)?

- Are there methods which are particularly appropriate for solving (1.2)?

- What can we say about the convergence of such a method?

- If we can establish convergence, what can we say about its asymptotic rate?

In this paper, we make preliminary attempts to answer all of these questions.

We will use the following notation. Bold lower and upper case Roman letters indicate vectors and matrices, respectively, while Greek and normal Roman letters denote scalars. Script style letters are index sets. A superscript (k) indicates a quantity which occurs at the k-th iteration or which is evaluated at $\boldsymbol{x}^{(k)}$.

We let $\boldsymbol{g}(\boldsymbol{x})$ and $\boldsymbol{H}(\boldsymbol{x})$, respectively, indicate the gradient, $\nabla_x f(\boldsymbol{x})$, and Hessian matrix, $\nabla_{xx} f(\boldsymbol{x})$, of the objective function. We define

$$(1.4) \qquad f_s^{(k)}(\boldsymbol{y}) \stackrel{\text{def}}{=} f(\boldsymbol{x}^{(k)} + \boldsymbol{S}^{(k)}\boldsymbol{y}),$$

$\boldsymbol{g}_s^{(k)}(\boldsymbol{y}) \stackrel{\text{def}}{=} \nabla_y f_s^{(k)}(\boldsymbol{y})$ and $\boldsymbol{H}_s^{(k)}(\boldsymbol{y}) \stackrel{\text{def}}{=} \nabla_{yy} f_s^{(k)}(\boldsymbol{y})$, and will make use of the derivative identities

$$(1.5) \qquad \boldsymbol{g}_s^{(k)}(\boldsymbol{y}) = \boldsymbol{S}^{(k)T}\boldsymbol{g}(\boldsymbol{x}^{(k)} + \boldsymbol{S}^{(k)}\boldsymbol{y})$$

and

$$(1.6) \qquad \boldsymbol{H}_s^{(k)}(\boldsymbol{y}) = \boldsymbol{S}^{(k)T}\boldsymbol{H}(\boldsymbol{x}^{(k)} + \boldsymbol{S}^{(k)}\boldsymbol{y})\boldsymbol{S}^{(k)}.$$

The paper is organised as follows. In Section 2, we consider the convergence of the algorithm given in the introduction. We discuss a number of ISM methods in Section 3, and we report on some preliminary numerical experience when solving some relatively large test examples, from the CUTE test suite (see Bongartz et al., 1995), in Section 4. Possible extensions, to the cases where there are linear or nonlinear constraints present, are given in our concluding Section 5, where we also offer our perspectives of this and future work.

2 Convergence of a general algorithm

Global convergence of the above scheme can be guaranteed under fairly general assumptions. Suppose that we are able to pick consecutive iterates $\boldsymbol{x}^{(k)}$ and $\boldsymbol{x}^{(k+1)} = \boldsymbol{x}^{(k)} + \boldsymbol{S}^{(k)}\boldsymbol{y}^{(k)}$ for which the Goldstein (1964) conditions

$$(2.1) \qquad f^{(k)} + \beta \boldsymbol{g}^{(k)T}\boldsymbol{S}^{(k)}\boldsymbol{y}^{(k)} \leq f^{(k+1)} \leq f^{(k)} + \alpha \boldsymbol{g}^{(k)T}\boldsymbol{S}^{(k)}\boldsymbol{y}^{(k)},$$

for some $0 < \alpha \leq \beta < 1$, are satisfied, where $\boldsymbol{y}^{(k)}$ is the approximate solution of (1.2). Suppose, furthermore, that

$$(2.2) \qquad \frac{-\boldsymbol{g}^{(k)T}\boldsymbol{S}^{(k)}\boldsymbol{y}^{(k)}}{\|\boldsymbol{S}^{(k)T}\boldsymbol{g}^{(k)}\|_2 \|\boldsymbol{y}^{(k)}\|_2} \geq \epsilon,$$

for some $\epsilon > 0$. Then the ISM algorithm from Section 1 is globally convergent to a stationary point for the problem (1.1) from any starting point so long as f is bounded from below and has a Lipschitz-continuous gradient (see, for example, Dennis and Schnabel, 1983, Theorem 6.3.3).

Using the definitions (1.4) and (1.5), we may write (2.1) as

$$(2.3) \qquad f_s^{(k)}(0) + \beta \boldsymbol{g}_s^{(k)}(0)^T \boldsymbol{y}^{(k)} \leq f_s^{(k)}(\boldsymbol{y}^{(k)}) \leq f_s^{(k)}(0) + \alpha \boldsymbol{g}_s^{(k)}(0)^T \boldsymbol{y}^{(k)}$$

and ensure that (2.2) is satisfied by requiring that

$$(2.4) \qquad \frac{-\boldsymbol{g}_s^{(k)}(0)^T \boldsymbol{y}^{(k)}}{\|\boldsymbol{g}_s^{(k)}(0)\|_2 \|\boldsymbol{y}^{(k)}\|_2} \geq \epsilon.$$

This is relevant as now the global convergence conditions may be verified in terms of the inner-minimization function f_s and its gradient. Similar global convergence results can be obtained if we replace condition (2.1) by the Armijo (1966) backtracking strategy (see Bertsekas, 1982, Section 1.3). It may, however, be difficult to design general algorithms which ensure that conditions (2.3) and (2.4) are satisfied on exit from the inner minimization. Thus, it may be preferable to impose extra conditions on the iterates generated during the inner minimization to ensure overall global convergence. With this in mind, suppose that we can find any point $y_s^{(k)}$ for which

$$f_s^{(k)}(0) + \beta g_s^{(k)}(0)^T y_s^{(k)} \leq f_s^{(k)}(y_s^{(k)}) \leq f_s^{(k)}(0) + \alpha g_s^{(k)}(0)^T y_s^{(k)} \tag{2.5}$$

and

$$\frac{-g_s^{(k)}(0)^T y_s^{(k)}}{\|g_s^{(k)}(0)\|_2 \|y_s^{(k)}\|_2} \geq \epsilon \tag{2.6}$$

are satisfied. Suppose, furthermore, that we terminate the inner minimization at a point $y^{(k)}$ for which

$$f_s^{(k)}(0) - f_s^{(k)}(y^{(k)}) \geq \tau(f_s^{(k)}(0) - f_s^{(k)}(y_s^{(k)})), \tag{2.7}$$

where $\tau > 0$. Then it is easy to show that this scheme is globally convergent under the same conditions as stated above. The advantage here is that the tests (2.5) and (2.6) need only be satisfied at an intermediate point to ensure convergence. Typically, the first inner-iterate provides such a point for carefully chosen subspaces and minimizers. For example, if the subspace contains the steepest-descent direction and the inner minimization starts by performing a linesearch in this direction, the resulting first inner-iterate satisfies (2.5) and (2.6). Similarly, if the subspace contains a (modified) truncated-Newton direction and the inner minimization starts by performing a linesearch in this direction, the same conclusion is true.

3 Computational Variants

We consider it important from a practical point of view to require that $S^{(k)}$ contains at least two components,

- a gradient-related direction, such as $-g^{(k)}$, to encourage global convergence, and
- a Newton-related direction, such as might be computed by a truncated-Newton method, to encourage fast asymptotic convergence, with safeguards to account for indefiniteness.

Of course, these components play a key role in dog-leg trust-region methods (see, for example, Powell, 1970). Additional components have the advantage of enlarging the subspace searched, but the disadvantage of increasing the overheads in solving the $s^{(k)}$-dimensional subspace minimization problem. In this section, we consider various possible ways of choosing the enlarged iterated subspace by choosing suitable conjugate directions.

3.1 Conjugate gradients

An appealing choice of $S^{(k)}$ may be obtained by picking the columns of $S^{(k)}$ as a set of $H^{(k)}$-conjugate directions, especially if these directions are generated by a conjugate-gradient (CG) method. In practise one would typically precondition first.

Suppose that $H^{(k)}$ is positive definite. Let $\phi^{(k)}(x^{(k)} + p)$ be the quadratic model,

$$\phi^{(k)}(x^{(k)} + p) = f^{(k)} + p^T g^{(k)} + \tfrac{1}{2} p^T H^{(k)} p, \tag{3.1}$$

of $f(\boldsymbol{x}^{(k)}+\boldsymbol{p})$ about $\boldsymbol{x}^{(k)}$. The preconditioned conjugate-gradient method (see, for example, Hestenes and Stiefel, 1952 and Golub and Loan, 1989, Section 10.3) is an iterative method which may be used to calculate the smallest value of (3.1). It is well known that the solution, $\boldsymbol{p}_n$, to this problem is the Newton direction.

A *preconditioner* $\boldsymbol{P}^{(k)}$ is usually an easily invertible approximation to $\boldsymbol{H}^{(k)}$. We shall insist that $\boldsymbol{P}^{(k)}$ has a uniformly bounded condition number. The j-th step of the preconditioned conjugate-gradient method determines the smallest value of (3.1) in the Krylov subspace spanned by the vectors $\{-((\boldsymbol{P}^{(k)})^{-1}\boldsymbol{H}^{(k)})^i(\boldsymbol{P}^{(k)})^{-1}\boldsymbol{g}^{(k)}\}_{i=0}^{j}$. Conjugacy properties ensure that each successive step may be accomplished by a univariate minimization of (3.1) in the direction $\boldsymbol{s}_j$; the vectors $\{\boldsymbol{s}_j\}$ are conjugate and are recurred from step to step. Significantly from our point of view, the first such vector, $\boldsymbol{s}_0 = -(\boldsymbol{P}^{(k)})^{-1}\boldsymbol{g}^{(k)}$. In exact arithmetic the method would terminate with the Newton direction, $\boldsymbol{p}_n$, after at most n steps, but numerical rounding errors ensure that the method behaves more like an infinite iteration (see Reid, 1971). Moreover, for the large-scale case, we would be unwilling to consider anywhere close to n iterations. Nonetheless, the method is an effective technique for calculating approximations to the Newton direction, especially if a good preconditioner is used or if low accuracy solutions may be tolerated (see Toint, 1981, Dembo et al., 1982, and Dembo and Steihaug, 1983).

In truncated-Newton methods, (see Dembo et al., 1982), the method of conjugate gradients is used to generate approximations to the Newton direction. The resulting search direction, $\boldsymbol{p}_{tn}$, is employed within a linesearch framework for solving unconstrained optimization problems. Highly accurate approximations to the Newton correction are only needed to accelerate the convergence of the iteration in the neighbourhood of a limit point, and crude improvements upon the steepest-descent direction suffice elsewhere. Furthermore, by monitoring the gradient of the model at each step of the conjugate-gradient method, we can decide when to terminate the iteration.

While such an approach has undoubtedly proved successful in practice, we note that a considerable amount of work is invested, in such a scheme, in calculating an "average" direction and that much of the information gleaned on the way is subsequently ignored. We take the point of view that directions generated by the conjugate-gradient method are of interest for the quadratic model, but might also be locally of interest for the true objective function. We thus propose to construct our iterated subspace from the subspace investigated by the conjugate gradient method.

We intend to include the following:

- The preconditioned steepest-descent direction, $\boldsymbol{s}_0 = -(\boldsymbol{P}^{(k)})^{-1}\boldsymbol{g}^{(k)}$;

- A number of other conjugate directions, $\boldsymbol{s}_j$, determined by the preconditioned conjugate-gradient method; and

- The overall truncated-Newton direction, $\boldsymbol{p}_{tn}$.

We note that the first of these components is designed to encourage global convergence, while the last will ensure that convergence occurs at a fast asymptotic rate. Thus, although we could include only the steepest descent direction and not Newton or some other combination, the choice above seems most prudent.

In the next two subsections, we will discuss the choice of the conjugate directions and the size of the iterated subspace.

3.2 Choice of conjugate directions

Suppose that, in addition to the (preconditioned) steepest-descent and (truncated) Newton directions, we wish to include q directions that are $\boldsymbol{H}^{(k)}$-conjugate in the subspace. Although there are other possibilities, we propose two specific ways to choose the conjugate directions. The simplest choice is just to take the first q generated (excluding, of course, the steepest-descent direction). Alternatively one might propose choosing the q which gave the largest decrease in the model. Experiments suggest that this is rarely more successful than the simpler scheme.

Another possibility is to include approximations from the extreme eigenspaces, that is the set of eigenvectors which correspond to the smallest and largest eigenvalues (recall $\boldsymbol{H}^{(k)}$ is assumed positive definite). In the unpreconditioned case, eigenvectors corresponding to large eigenvalues correspond to the directions along which the function and CG model change most rapidly. In the preconditioned case, they still have the same effect for the transformed model.

Similarly, those associated with small eigenvalues reflect the (transformed) space in which the Newton direction is likely to be sensitive and contributions from this space are necessary if rapid progress is to be made. Thus both sets of vectors are reasonable candidates for subspace directions.

Clearly, the calculation of these spaces is generally prohibitively expensive, but they may be approximated by directions generated during the CG process (see, eg, Parlett, 1980, Chapter 13). One possibility is to monitor the Rayleigh quotients

$$(3.2) \qquad \frac{\boldsymbol{s}_j^T \boldsymbol{H}^{(k)} \boldsymbol{s}_j}{\boldsymbol{s}_j^T \boldsymbol{s}_j}$$

and include the $\boldsymbol{s}_j$ which give rise to the most extreme Rayleigh quotients. Of course, these vectors are not eigenvectors of $\boldsymbol{H}^{(k)}$, but they usually contain significant contributions in the extreme eigenspaces.

3.3 Choice of subspace dimension

The choice of subspace dimension is clearly important. The simplest choice is to fix an upper bound s on this dimension before the computation proceeds (perhaps $s = 10$, see Section 4), and to select $s^{(k)}$ to be the smaller of s and the total number of CG directions sampled during the k-th CG iteration — recall that the CG process may be truncated and thus fewer than s directions may have been computed.

A more sophisticated approach is to try to dynamically select the size of subspace based upon the needs of the k-th iteration. For instance, if the Hessian is relatively well-conditioned, it is reasonable that a subspace made up from the steepest-descent and (truncated) Newton directions will suffice. If, on the other hand, the Hessian is ill-conditioned, further subspace directions to account for the extreme eigenvalues are likely to prove beneficial.

A simple heuristic would be to monitor the Rayleigh quotient as the CG iteration proceeds. Typically the first search direction, $\boldsymbol{s}_0 = -(\boldsymbol{P}^{(k)})^{-1}\boldsymbol{g}^{(k)}$, for the CG iteration will contain components of all eigenvectors and hence some components of those corresponding to the largest eigenvalues. The influence of the eigenvectors corresponding to the large eigenvalues is reduced in the subsequent directions $\boldsymbol{s}_1, \ldots$, and this is reflected in a reduction in the Rayleigh quotient during these iterations. This effect is reversed after a number of iterations, when the influence of the larger eigenvalues reappears. It would thus seem sensible to record the iteration number, $i^{(k)}$, at which the Rayleigh quotient first starts to increase after its initial sequence of decreases. As we know that we then have sampled eigenvectors corresponding to both "large" and "small" eigenvalues, it is appropriate to set $s^{(k)} = i^{(k)}$.

3.4 The inner minimization

As we have stated, we believe that there are a number of highly effective algorithms for the unconstrained minimization of a function of several variables. Indeed, it would not be unreasonable to say that the problem has effectively been solved so long as derivatives are available. Among the most successful methods are the Newton-like second-derivative methods and the finite-difference and secant methods which require only gradients (see, for example, Dennis and Schnabel, 1983, Gill et al., 1981 or Fletcher, 1987).

When considering the minimization of (1.4), we note that the calculation of derivatives of $f_s^{(k)}$ requires those of f. We see that the calculation of the second derivatives (1.6) requires significantly more products involving $S^{(k)}$ than do the first derivatives (1.5). Thus, we would prefer to use methods which either only require relatively few Hessian-vector products, such as (preconditioned and truncated) conjugate-gradient methods, or secant methods, which build up approximations to the second derivatives from gradients in the $s^{(k)}$-dimensional subspace as they proceed.

The most widely used secant methods are those in the convex Broyden class of positive-definite approximations, of which the BFGS method has the best reputation (again see, for example, Dennis and Schnabel, 1983). While there is some controversy as to whether there are better nonconvex secant updates (see for example, Conn et al., 1991, Khalfan et al., 1993, and Byrd et al., 1996), we feel that convex secant methods are most natural in a linesearch, as opposed to a trust-region, context. Such methods start with a positive-definite second-derivative approximation and generate a sequence of matrices which mimic the curvature in the space of directions searched. Traditionally, the Cholesky factors of the sequence are updated as the iteration proceeds. We now show that building a good starting matrix in our case is easy.

Firstly, suppose that $S^{(k)}$ is made up purely of $H^{(k)}$-conjugate directions. Then the exact second-derivative matrix $H_s^{(k)}$ is diagonal because of the conjugacy and moreover its diagonal entries will have been calculated during the conjugate-gradient process. This matrix, then, is a good starting approximation; its Cholesky factors are trivial to determine. Furthermore, this choice ensures that the first quasi-Newton search direction is identical to that generated by minimizing the quadratic model (3.1) in the manifold $x^{(k)} + S^{(k)}y$ (see, Gill et al., 1981, section 4.8.3.1).

Now suppose that $S^{(k)}$ is made by augmenting a set of $H^{(k)}$-conjugate directions by the overall truncated-Newton direction p_{tn}. Then, the exact second-derivative matrix has an arrowhead structure with the leading $s^{(k)} - 1$ by $s^{(k)} - 1$ submatrix being diagonal and the remaining row and column easy to obtain. To be precise, if we denote the residual $H^{(k)}p_{tn} + g^{(k)}$ following the truncated conjugate-gradient process by $r^{(k)}$, the last column of the required second-derivative matrix is $S^{(k)T}(r^{(k)} - g^{(k)})$. Thus, once again this matrix provides a good starting approximation in that its Cholesky factors are extremely cheap to compute. Moreover, as before, the first quasi-Newton direction gives the minimum of the model (3.1) in the manifold $x^{(k)} + S^{(k)}y$. Significantly, as the truncated-Newton direction is in the subspace, this first quasi-Newton direction is thus the same as the truncated-Newton direction.

4 Numerical Experiments

We start this section by investigating, from a numerical point of view, the impact of different subspace sizes on the convergence of the method. As the problem **FMINSURF**[1] is particularly efficiently solved using ISM in comparison with the default version of **LANCELOT**, we examined this problem in detail. We ran ISM, without preconditioner, but constructing the subspace from a combination of the steepest-descent, truncated-Newton and extreme directions as described in Section 3.2, using a variety of subspace dimensions and illustrate, in Figure 1, the effects of the choice of this dimension on the CPU time[2] required to solve the problem.

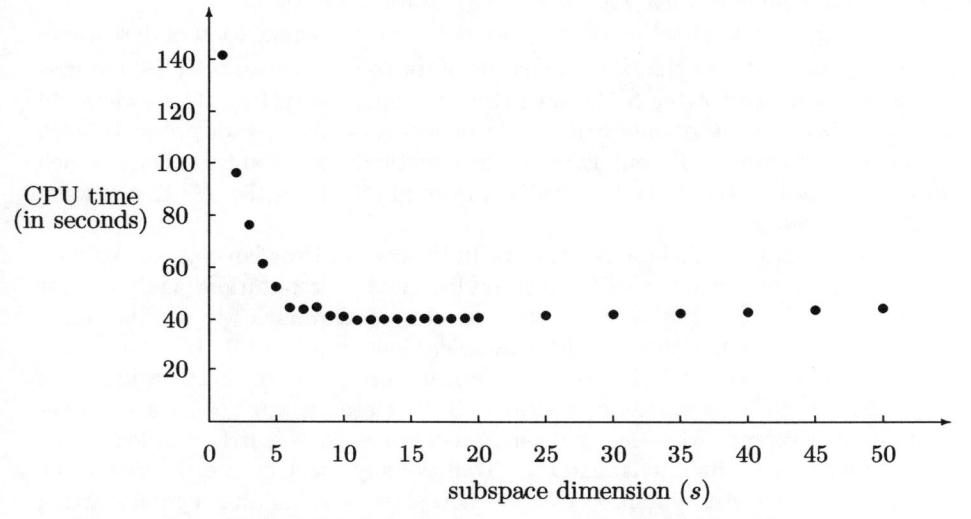

FIG. 1. *The impact of varying the subspace dimension on* **FMINSURF** *(using an unpreconditioned ISM in which the subspace is chosen from the the steepest-descent, truncated-Newton and extreme CG directions).*

We observe that the CPU time for small subspace dimension is large, but that, for subspaces of dimension between nine and forty, the time is relatively constant, being within ten percent of the least ($s = 11$) time. Thus, it appears that for this problem, adding information above the steepest-descent and truncated-Newton directions is beneficial but there is little extra payoff from using more than eleven directions. When we monitored the Rayleigh quotients for this problem, we observed that the quotient decreases for on average ten CG iterations before increasing for the first time. Thus, unless we insist on at least ten CG iterations, it is possible that we may not have sampled the complete eigenspace. Similar runs on different problems indicate that this behaviour is not exceptional. In other words it appears to be not exceptional that searching in a subspace that uses eight directions

---

[1] From the CUTE collection, see Bongartz *et al.* (1995).
[2] This experiment was performed on an IBM RISC/6000 320h workstation, using optimized (-O) Fortran 77 code and IBM-supplied BLAS.

corresponding to the extreme Rayleigh quotients with the addition of the steepest descent and the truncated Newton directions captures the bulk of the possible improvements at a given iteration. Of course, this cannot be the case for all iterations or all problems, but the overhead associated with choosing substantially more directions, and the loss of descent incurred by using less directions (recall s is fixed), does not payoff in general.

We next report the results of running a number of variants of our Iterated-Subspace Minimization code, based on the suggestions made in Section 3, on some large or difficult test problems. The simplest variant has the following features.

- *No* preconditioning is used.

- $s^{(k)}$ is the smaller of the *fixed* bound $s = 10$ (as suggested by the above example) and the number of inner iterations required to determine the truncated-Newton direction.

- The subspace is constructed from the *first* $s^{(k)} - 1$ conjugate directions plus the truncated-Newton direction. The truncation is performed when the residual (gradient) of the model is smaller than $\|g^{(k)}\|_2 \min(0.1, \|g^{(k)}\|_2^{0.1})$ or when more than n conjugate-gradient iterations were performed.

- The model is modified, if necessary, to ensure that it is strictly convex. The modification is carried out as the conjugate-gradient iteration proceeds using the method of Arioli *et al.* (1993).

- The iteration is deemed to have converged when $\|g^{(k)}\|_2$ is smaller than 10^{-5}.

- A BFGS linesearch method is used to solve the inner-minimization problem. An Armijo backtracking linesearch is used, starting with a unit step and dividing the step by two until the Armijo sufficient decrease condition is satisfied. If a step of one proves acceptable, but the model has been modified to ensure that it is strictly convex, the step is doubled until an unacceptable stepsize is determined, when the last-found acceptable step is chosen. A maximum of $2s^{(k)}$ BFGS iterations are permitted and the iteration is stopped if $\|g_s^{(k)}\|_2$ is smaller than 10^{-6}.

Note that one Hessian evaluation is made for each major iteration and one gradient evaluation for each inner iteration. We denote this method by the symbol ISM(n,f,f), where n means that *no* preconditioning is used, the first f that the subspace is constructed from the *first* CG directions, and the second f that the maximal subspace dimension s is *fixed* (to ten).

We also consider the ISM(p,f,f) method, which is identical to ISM(n,f,f), except that an 11-band modified Cholesky factorization *preconditioner* is used in the conjugate-gradient calculation. The factorization takes the elements of $H^{(k)}$ within a band of semi-bandwidth five of the diagonal, replacing any other elements by zeros. The resulting band matrix is factorized, modifications being made according to the method of Schnabel and Eskow (1991) to ensure that the preconditioner is positive definite with bounded condition number.

We define the methods ISM(n,e,f) and ISM(p,e,f) as the modifications of ISM(n,f,f) and ISM(p,f,f), respectively, where we construct the subspace from a set of *extreme* $s^{(k)} - 2$ conjugate directions plus the steepest-descent and truncated-Newton directions. We pick half of the extreme directions to be those whose Rayleigh quotient is largest, while the remainder correspond to the smallest Rayleigh quotients.

We next consider the possibility of generating the subspace dimension *automatically*, as discussed in Section 3.3. Once the subspace dimension has been determined, we construct the subspace from the set of first or extreme $s^{(k)}-2$ conjugate directions plus the steepest-descent and truncated-Newton direction, just as before. This results in four additional variants, namely ISM(n,f,a), ISM(p,f,a), ISM(n,e,a) and ISM(p,e,a).

As a yard-stick, we compare the above methods with three other algorithms. The first is the default version of the LANCELOT A nonlinear optimization package (see Conn et al., 1992) (denoted LAN(p)) in which an 11-band preconditioner is used, together with the same unpreconditioned version (denoted LAN(n)). LANCELOT is a trust-region method in which a gradient and Hessian evaluation are made on every successful iteration. The trust region subproblem is solved using a truncated conjugate gradient method. The second algorithm included in our comparison is a truncated-Newton method (see Dembo and Steihaug (1983)). The truncated-Newton search direction is obtained by an inexact minimization of the Newton model using unpreconditioned or preconditioned conjugate gradients. We denote the resulting methods by TN(n) and TN(p), respectively. These two variants are obtained from our ISM algorithms by restricting the subspace minimization to a single linesearch along the truncated-Newton direction. Here, the truncation is performed when the residual (gradient) of the model is smaller than $\|g^{(k)}\|_2 \min(0.1, \|g^{(k)}\|_2^{0.5})$ or when more than n conjugate-gradient iterations are performed[3]. Finally, we also compare our ISM algorithms with the limited memory algorithm L-BFGS-B of Zhu et al. (1996) (see also Byrd et al., 1995). In the unconstrained case, Algorithm L-BFGS-B approximately minimizes a quadratic model, whose Hessian is a limited memory BFGS approximation of the Hessian of the objective function, to compute a search direction, and performs a linesearch along this search direction. We set parameter **isbmin** to three in the code, meaning that the conjugate-gradient method is used to compute the search direction. As no preconditioning is considered in this method (denoted LM(n)), we compare it with the unpreconditioned variants of the ISM algorithms. We note that all the algorithms considered in this comparison, except the limited memory one of course, use exact first and second derivatives.

We selected our 34 test examples as the majority of large and/or difficult unconstrained test examples in the CUTE (see Bongartz et al., 1995) test set. Only problems which took excessive CPU time (more than 30 minutes), or which were variations on the reported problems, were excluded. All experiments were made on a DEC 3000 workstation, using optimized (-O) Fortran 77 code.

Tables 1 and 2 first report cumulative statistics on the performance of the considered algorithms on 32 problems for the unpreconditioned case (the limited memory algorithm failed to solve problems DIXON3DQ and VARDIM), and on 32 problems for the preconditioned case (different approximate local optima have been reached for problems LIARWHD and NONDIA). In these tables and the following ones, #f indicates the total number of function evaluations and *time* the total CPU time (in seconds). The second column indicates which of the tables in the appendix gives the complete and detailed results for the considered method.

These tables show some interesting results. In particular, they indicate that the automatic choice of the subspace dimension is advantageous on average. On the other hand, there is little difference between the average performance of ISM methods using the first CG directions to define the subspace and those using the extreme ones. One also sees that, on average, the preconditioned ISM variants are all better in CPU time

---

[3]The exponent differs from the 0.1 used for the ISM methods because otherwise the results for the truncated-Newton method deteriorate considerably.

TABLE 1
Cumulative statistics on the performance of all methods on 32 problems (unpreconditioned case)

Method	Details	#f	Time
LAN(n)	Table 5	5585	486
TN(n)	Table 7	19443	649
LM(n)	Table 9	15046	1175
ISM(n,f,f)	Table 10	21743	695
ISM(n,e,f)	Table 12	22168	720
ISM(n,f,a)	Table 14	14753	532
ISM(n,e,a)	Table 16	15172	533

TABLE 2
Cumulative statistics on the performance of all except the limited memory method on 32 problems (preconditioned case)

Method	Details	#f	Time
LAN(p)	Table 6	4197	456
TN(p)	Table 8	7852	471
ISM(p,f,f)	Table 11	22096	649
ISM(p,e,f)	Table 13	22193	652
ISM(p,f,a)	Table 15	12616	448
ISM(p,e,a)	Table 17	12774	458

than their unpreconditioned counterparts. When comparing with the other methods, the unpreconditioned ISM variants with automatic choice of the subspace dimension require, on average, more function evaluations than **LANCELOT**, but approximately the same amount as limited memory and less than truncated-Newton. In the preconditioned case, the total number of function evaluations is less for **LANCELOT** and truncated-Newton. The overall CPU time for the unpreconditioned case indicates a better performance, on average, for **LANCELOT**, directly followed (within less than ten per cent) by ISM variants with automatic choice of the subspace dimension. The limited memory method did not perform as well on these examples. This is more than likely because it uses less derivative information. When preconditioning is used, the total CPU time for the ISM methods with automatic choice of the subspace dimension is comparable with (and even slightly better than) the total CPU time for **LANCELOT** and for truncated-Newton.

We next consider a more disaggregate presentation of what happens problem by problem to refine our analysis. We present in Tables 3 and 4 the number of times that each of the considered method ranks first, second, third, etc. for the two criteria used above. In these rankings, two CPU times are reputed identical if they differ by less than five percent or by less than half a second.

These tables strengthen the indications drawn above from average measures, except that they indicate that the supremacy of **LANCELOT** in CPU time for the unpreconditioned ISM variants is misleading. It appears indeed that all the ISM methods (with or without preconditioner) are very competitive in CPU time. We further observe that truncated-Newton is sometimes very good and sometimes rather poor for the unpreconditioned case (as may be expected), while this dichotomy is less marked in the preconditioned case.

TABLE 3
Rankings (unpreconditioned case)

Method	Function evaluations (#f)							Time						
	1st	2nd	3rd	4th	5th	6th	7th	1st	2nd	3rd	4th	5th	6th	7th
LAN(n)	26	2	1	1	0	2	0	12	3	1	1	3	6	6
TN(n)	3	9	4	3	0	0	13	18	0	0	1	1	0	12
LM(n)	1	10	1	2	4	5	9	14	1	1	0	5	5	6
ISM(n,f,f)	1	6	6	3	5	9	2	19	0	1	3	4	2	3
ISM(n,e,f)	1	6	5	3	5	8	4	19	1	1	1	2	7	1
ISM(n,f,a)	3	9	7	9	4	0	0	22	1	2	4	1	2	0
ISM(n,e,a)	3	6	9	8	4	2	0	21	0	4	2	4	1	0

TABLE 4
Rankings (preconditioned case)

Method	Function evaluations (#f)						Time					
	1st	2nd	3rd	4th	5th	6th	1st	2nd	3rd	4th	5th	6th
LAN(p)	31	0	0	0	0	1	19	2	0	4	0	7
TN(p)	5	13	0	10	0	4	20	0	0	2	3	7
ISM(p,f,f)	5	2	3	5	16	1	17	2	1	3	7	2
ISM(p,e,f)	5	2	4	3	15	3	15	0	3	3	9	2
ISM(p,f,a)	5	13	9	1	3	1	19	4	6	2	0	1
ISM(p,e,a)	6	8	11	2	5	0	19	2	4	4	3	0

We thus conclude that in general ISM methods are efficient from the CPU time point of view, even though they may require substantially more function evaluations than LANCELOT (but not more than truncated-Newton and limited memory). Further, the automatic determination of the subspace dimension is often quite helpful, both in the unpreconditioned and the preconditioned cases. On the other hand, there is little difference in performance between ISM variants that build the subspace from the first CG directions and variants that use the extreme ones. As the former is easier to implement, one might prefer it in practice. In fact ISM(p,f,a) is the best, slightly[4].

5 Perspectives and Conclusions

In this paper, we have shown that it is possible to solve large-scale nonlinear optimization problems using methods designed for small-scale problems. These methods may be regarded as a generalization of linesearch-type methods more usually used to solve unconstrained minimization problems. We have indicated that the convergence of our methods depends upon using robust algorithms for the small-dimensional subproblems, and have suggested a number of ways of selecting promising subspaces in which to search. Furthermore, we feel that there are a number of important areas for future investigation.

---

[4] One should also bear in mind, at this point, that LANCELOT is a much more sophisticated code than our ISM variants, because it is designed for solving generally constrained in addition to unconstrained problems, has been extensively tested and refined, and contains a number of safeguards that are not included in the simpler ISM codes.

- It is possible to extend the iterated-subspace minimization idea to treat linearly constrained optimization problems. For example, if we suppose the original problem is of the form

  $$\text{(5.1)} \qquad \underset{x \in \Re^n}{\text{minimize}} \ \ f(x) \ \ \text{subject to} \ \ l \leq Ax \leq u,$$

  where A is an m by n matrix, l and u are m-vectors and $x^{(k)}$ satisfies $l \leq Ax^{(k)} \leq u$, we may apply the following linearly-constrained ISM algorithm:
  1. Stop if convergence tests are satisfied.
  2. Determine a full-rank subspace matrix $S^{(k)} \in \Re^{n \times s^{(k)}}$, where $s^{(k)} \ll n$.
  3. Approximately solve the $s^{(k)}$-dimensional minimization problem

  $$\text{(5.2)} \qquad \underset{y \in \Re^{s^{(k)}}}{\text{minimize}} \ \ f(x^{(k)} + S^{(k)}y) \ \ \text{subject to} \ \ l^{(k)} \leq A^{(k)}y \leq u^{(k)},$$

  where $A^{(k)} = AS^{(k)}$, $l^{(k)} = l - Ax^{(k)}$ and $u^{(k)} = u - Ax^{(k)}$, and set

  $$\text{(5.3)} \qquad x^{(k+1)} = \text{(approximate)} \ \underset{y \in \Re^{s^{(k)}}}{\arg \min} \ f(x^{(k)} + S^{(k)}y)$$

  $$\text{subject to} \ \ l^{(k)} \leq A^{(k)}y \leq u^{(k)}.$$

  The central issues remain those discussed in Section 1. However, extra care must be exercised when picking the subspace matrix, as it is now desirable for a *constrained* steepest-descent and (truncated) Newton directions to lie in the subspace. The real difficulty is in adding appropriate additional directions since, for example, a straightforward analogue of the unconstrained case would require that we have conjugate directions in the constrained (reduced) space. One instance where this supposition is reasonable is when the constraints are just simple bounds. We also now need to use efficient methods for solving *small linearly-constrained minimization* problems when determining $x^{(k+1)}$, but fortunately the state-of-the-art here is as advanced as it is for unconstrained minimization.

  The ISM philosophy does not obviously extend to handle nonlinearly-constrained minimization problems except that, of course, any unconstrained or linearly-constrained subproblems may be treated by existing ISM methods. This may be important for nonlinearly-constrained minimization methods which are based on the sequential minimization of penalty or barrier functions, or their augmented or shifted counterparts.

- In our investigations, we found it convenient to use a linesearch (BFGS) method to solve the inner-iteration subproblems. One might, of course, alternatively use a trust-region method to solve the subproblem. However, as the performance of such methods depends upon building a good model within an adequate trust region, and as our ISM method will solve a sequence of subproblems, it may be that a good trust-region radius for one subproblem is poor for the next, and inefficiencies may occur. Thus care may be needed in determining interactions between successive models.

Another important issue is how to pick stopping rules, analogous to (2.5)–(2.7), which are appropriate for trust-region based methods. The main difficulty here is that the initial trust-region radius may interfere with a condition like (2.5).

- While we have suggested a number of methods for computing a good iterated subspace, more work clearly needs to be performed. We believe that we have identified some of the ingredients of a good subspace, but our understanding is far from complete.

6 Acknowledgement

Nick Gould would like to thank CERFACS for the facilities which made some of this research possible.

References

[Arioli et al., 1993] M. Arioli, T. F. Chan, I. S. Duff, N. I. M. Gould, and J. K. Reid. Computing a search direction for large-scale linearly constrained nonlinear optimization calculations. Technical Report Research Report 93-066, RAL, Chilton, England, 1993.

[Armijo, 1966] L. Armijo. Minimization of functions having Lipschitz-continuous first partial derivatives. *Pacific Journal of Mathematics*, 16:1–3, 1966.

[Bertsekas, 1982] D. P. Bertsekas. *Constrained Optimization and Lagrange Multipliers Methods*. Academic Press, London, 1982.

[Bongartz et al., 1995] I. Bongartz, A. R. Conn, N. I. M. Gould, and Ph. L. Toint. CUTE: Constrained and unconstrained testing environment. *ACM Transactions on Mathematical Software*, 21(1):123 – 160, 1995.

[Byrd et al., 1995] R. H. Byrd, P. Lu, J. Nocedal, and C. Zhu. A limited memory algorithm for bound constrained optimization. *SIAM Journal on Scientific Computing*, 16(5):1190–1208, 1995.

[Byrd et al., 1996] R. H. Byrd, H.F. Khalfan, and R. B. Schnabel. Analysis of a symmetric rank-jone trust region method. *SIAM Journal on Optimization*, 1996. to appear.

[Conn et al., 1991] A. R. Conn, N. I. M. Gould, and Ph. L. Toint. Convergence of quasi-Newton matrices generated by the symmetric rank one update. *Mathematical Programming*, 50(2):177–196, 1991.

[Conn et al., 1992] A. R. Conn, N. I. M. Gould, and Ph. L. Toint. LANCELOT: *a Fortran package for large-scale nonlinear optimization (Release A)*, volume 17 of *Springer Series in Computational Mathematics*. Springer Verlag, Heidelberg, Berlin, New York, 1992.

[Conn et al., 1993] A. R. Conn, N. I. M. Gould, and Ph. L. Toint. Complete numerical results for LANCELOT Release A. Research Report RC 18750, IBM T. J. Watson Research Center, Yorktown Heights, NY 10598, USA, 1993. Also issued as Technical Report 92/15, Department of Mathematics, FUNDP, Namur, Belgium.

[Conn et al., 1996] A. R. Conn, N. I. M. Gould, and Ph. L. Toint. Numerical experiments with the LANCELOT package (Release A) for large-scale nonlinear optimization. *Mathematical Programming*, to appear, 1996.

[Cragg and Levy, 1969] E. E. Cragg and A. V. Levy. Study on a supermemory gradient method for the minimization of functions. *Journal of Optimization Theory and Applications*, 4(3):191–205, 1969.

[Dembo and Steihaug, 1983] R. S. Dembo and T. Steihaug. Truncated-Newton algorithms for large-scale unconstrained optimization. *Mathematical Programming*, 26:190–212, 1983.

[Dembo et al., 1982] R. S. Dembo, S. C. Eisenstat, and T. Steihaug. Inexact-Newton methods. *SIAM Journal on Numerical Analysis*, 19(2):400–408, 1982.

[Dennis and Schnabel, 1983] J. E. Dennis and R. B. Schnabel. *Numerical methods for unconstrained optimization and nonlinear equations*. Prentice-Hall, Englewood Cliffs, NJ, 1983.

[Dennis and Turner, 1987] J. E. Dennis and K. Turner. Generalized conjugate directions. *Linear Algebra and Applications*, 88/89:187–209, 1987.

[Dixon et al., 1984] L. C. W. Dixon, K. D. Patel, and P. G. Ducksbury. Experience running optimization algorithms on parallel processing system. In *11th IFIP Conference on System Modelling and Optimization*, volume 59, pages 891–932, Berlin, 1984. Springer Verlag. Lecture Notes in Control and Information Sciences.

[Fletcher, 1987] R. Fletcher. *Practical Methods of Optimization*. J. Wiley and Sons, Chichester, second edition, 1987.

[Gill et al., 1981] P. E. Gill, W. Murray, and M. H. Wright. *Practical Optimization*. Academic Press, London and New York, 1981.

[Goldstein, 1964] A.A. Goldstein. Convex programming in Hilbert space. *Bulletin of the American Mathematical Society*, 70:709–710, 1964.

[Golub and Loan, 1989] G. H. Golub and C. F. Van Loan. *Matrix Computations*. Johns Hopkins University Press, Baltimore, second edition, 1989.

[Hestenes and Stiefel, 1952] M. R. Hestenes and E. Stiefel. Methods of conjugate gradients for solving linear systems. *J. Res. N.B.S.*, 49:409–436, 1952.

[Kaporin and Axelsson, 1995] I. E. Kaporin and O. Axelsson. On a class of nonlinear equation solvers based on the residual norm reduction over a sequence of affine subspaces. *SIAM Journal on Scientific Computing*, 16(1):228–249, 1995.

[Khalfan et al., 1993] H. Khalfan, R. H. Byrd, and R. B. Schnabel. The symmetric rank one update. Technical Report CU-CS-240, Department of Computer Science, University of Colorado, Boulder, USA", 1993.

[Miele and Cantrell, 1969] A. Miele and J. W. Cantrell. Study on a memory gradient method for the minimization of functions. *Journal of Optimization Theory and Applications*, 3:459–470, 1969.

[Parlett, 1980] B. N. Parlett. *The Symmetric Eigenvalue Problem*. Prentice-Hall, Englewood Cliffs, USA, 1980.

[Powell, 1970] M. J. D. Powell. A hybrid method for nonlinear equations. In P. Rabinowitz, editor, *Numerical Method for Nonlinear Algebraic Equations*, pages 87–114. Gordon and Breach, London, 1970.

[Reid, 1971] J. K. Reid. On the method of conjugate gradients for the solution large sparse linear equations. In J. K. Reid, editor, *Large sparse sets of linear equations*, pages 231–254, London, 1971. Academic Press.

[Saad, 1990] Y. Saad. Krylov subspace methods: theory, algorithms and applications. In *Computational Methods in Applied Science and Engineering, INRIA 29 January - 2 February*. INRIA, 1990.

[Schnabel and Eskow, 1991] R. B. Schnabel and E. Eskow. A new modified Cholesky factorization. *SIAM Journal on Scientific and Statistical Computing*, 11:1136–1158, 1991.

[Toint, 1981] Ph. L. Toint. Towards an efficient sparsity exploiting Newton method for minimization. In I. S. Duff, editor, *Sparse Matrices and Their Uses*, pages 57–88. Academic Press, London, 1981.

[Vinsome, 1976] P. K. W. Vinsome. Orthomin, an iterative method for solving sparse sets of simultaneous linear equations. In *Proceedings of the Fourth Symposium on Reservoir Simulation*. Society of Petroleum Engineers of AIME, 1976.

[Zhu et al., 1996] C. Zhu, R. H. Byrd, P. Lu, and J. Nocedal. L-BFGS-B: Fortran subroutines for large-scale bound constrained optimization. *ACM Transactions on Mathematical Software*, (to appear), 1996.

A Detailed numerical results

In this appendix, we give comprehensive details of the performance of each of the methods discussed in the main body of the paper on the complete set of test examples. We include these results so that others may, in future, compare new proposals with ours.

TABLE 5
Results for LAN(n) *(unpreconditioned* LANCELOT*) on large or hard problems.*

problem	n	#f	#it	#cg	time	f
ARWHEAD	1000	6	5	2	0.95	1.6903D-10
BDQRTIC	1000	13	12	66	2.01	3.9838D+03
BROWNAL	100	3	2	3	0.17	1.1263D-13
BRYBND	1000	14	13	177	3.53	2.8797D-13
CRAGGLVY	1000	14	13	166	2.40	3.3642D+02
DIXMAANA	1500	7	6	10	1.40	1.0000D+00
DIXON3DQ	1000	6	5	1427	5.04	7.7032D-09
DQDRTIC	1000	3	2	0	0.73	1.6602D-23
DQRTIC	1000	36	35	223	2.87	3.7687D-06
EDENSCH	1000	15	14	35	1.58	6.0033D+03
EIGENALS	110	16	15	142	0.76	2.7564D-11
ENGVAL1	1000	8	7	20	1.08	1.1082D+03
FLETCHCR	1000	3529	3528	20619	219.85	5.0581D-14
FMINSURF	1024	190	189	527	15.69	1.0000D+00
FREUROTH	1000	21	20	40	1.76	1.2147D+05
GENROSE	1000	1214	1213	6640	77.43	1.0000D+00
LIARWHD	1000	14	13	14	1.27	2.3011D-16
MANCINO	100	12	11	13	11.28	4.3367D-17
MOREBV	1000	3	2	265	2.01	2.0186D-09
NCB20B	1000	28	27	1274	96.53	1.6760D+03
NONDIA	1000	30	29	29	1.99	4.3483D-16
NONDQUAR	1000	92	91	853	6.43	9.5606D-06
PENALTY1	1000	56	55	44	8.15	9.6862D-03
POWELLSG	1000	16	15	60	1.35	1.1167D-06
POWER	1000	29	28	613	3.73	5.8399D-09
QUARTC	1000	36	35	223	3.07	3.7687D-06
SINQUAD	1000	70	69	126	5.54	6.3238D-05
SROSENBR	1000	11	10	20	1.06	6.6723D-12
TOINTGSS	1000	9	8	15	0.94	1.0010D+01
TQUARTIC	1000	14	13	14	1.26	1.0168D-16
TRIDIA	1000	5	4	91	1.14	8.5487D-14
VARDIM	1000	37	36	0	1.87	1.1203D-20
VAREIGVL	1000	13	12	313	4.32	3.0691D-08
WOODS	1000	58	57	194	3.42	3.5419D-16

Key: n = number of variables, $\#f$ = number of function evaluations, $\#it$ = number of iterations, $\#cg$ = total number of CG iterations, $time$ = total CPU time in seconds, f = smallest function value obtained.

TABLE 6
Results for LAN(p) *(default preconditioned* LANCELOT*) on large or hard problems.*

problem	n	#f	#it	#cg	time	f
ARWHEAD	1000	6	5	1	0.72	1.6903D-10
BDQRTIC	1000	12	11	13	1.56	3.9838D+03
BROWNAL	100	4	3	40	0.97	6.3470D-11
BRYBND	1000	17	16	30	2.53	1.8775D-12
CRAGGLVY	1000	15	14	11	1.51	3.3642D+02
DIXMAANA	1500	8	7	10	1.14	1.0000D+00
DIXON3DQ	1000	3	2	2	0.40	0.0000D+00
DQDRTIC	1000	3	2	0	0.40	1.6602D-23
DQRTIC	1000	36	35	27	2.31	3.6952D-06
EDENSCH	1000	13	12	9	1.24	6.0033D+03
EIGENALS	110	25	24	56	1.18	1.4887D-12
ENGVAL1	1000	8	7	7	0.81	1.1082D+03
FLETCHCR	1000	2138	2137	2136	119.01	2.8160D-12
FMINSURF	1024	316	315	436	106.32	1.0000D+00
FREUROTH	1000	11	10	7	1.06	1.2147D+05
GENROSE	1000	1095	1094	1158	63.37	1.0000D+00
LIARWHD	1000	15	14	21	1.29	2.0274D-20
MANCINO	100	16	15	8	17.39	9.2449D-18
MOREBV	1000	2	1	1	0.39	7.3289D-13
NCB20B	1000	23	22	568	50.35	1.6760D+03
NONDIA	1000	30	29	47	2.18	6.9910D-16
NONDQUAR	1000	18	17	19	1.21	1.3932D-09
PENALTY1	1000	64	63	432	23.16	9.6862D-03
POWELLSG	1000	16	15	15	1.06	2.2324D-06
POWER	1000	28	27	53	7.60	1.8888D-08
QUARTC	1000	36	35	27	2.32	3.6952D-06
SINQUAD	1000	132	131	341	14.82	9.0365D-07
SROSENBR	1000	11	10	10	0.82	5.8309D-13
TOINTGSS	1000	3	2	2	0.44	1.0000D+01
TQUARTIC	1000	13	12	24	1.19	2.9487D-11
TRIDIA	1000	3	2	1	0.49	1.3827D-32
VARDIM	1000	37	36	0	19.03	1.1203D-20
VAREIGVL	1000	13	12	199	7.06	4.1621D-08
WOODS	1000	72	71	69	4.18	2.4231D-11

Key: n = number of variables, $\#f$ = number of function evaluations, $\#it$ = number of iterations, $\#cg$ = total number of CG iterations, *time* = total CPU time in seconds, f = smallest function value obtained.

TABLE 7
Results for the unpreconditioned truncated-Newton method TN(n) *on large or hard problems.*

problem	n	#f	#its	#cg	time	f
ARWHEAD	1000	1545	389	395	26.63	0.00D+00
BDQRTIC	1000	43	17	97	1.68	3.98D+03
BROWNAL	100	5	3	4	0.04	2.45D-08
BRYBND	1000	37	14	48	1.49	1.22D-12
CRAGGLVY	1000	45	16	129	1.98	3.36D+02
DIXMAANA	1500	189	64	73	7.10	1.00D+00
DIXON3DQ	1000	7	4	1748	3.91	1.25D-11
DQDRTIC	1000	10	5	12	0.23	1.91D-14
DQRTIC	1000	48	17	70	0.38	5.14D-08
EDENSCH	1000	46	17	44	1.35	6.00D+03
EIGENALS	110	69	22	179	0.98	8.43D-12
ENGVAL1	1000	31	13	28	0.81	1.11D+03
FLETCHCR	1000	3046	1184	13315	100.03	3.04D-15
FMINSURF	1024	156	26	5559	46.24	1.00D+00
FREUROTH	1000	92	19	110	2.42	1.21D+05
GENROSE	1000	2896	540	9838	71.28	1.00D+00
LIARWHD	1000	78	28	32	1.42	1.43D-16
MANCINO	100	72	21	29	68.27	1.87D-21
MOREBV	1000	3	2	549	2.14	1.34D-09
NCB20B	1000	74	19	1053	81.98	1.68D+03
NONDIA	1000	156	55	58	3.46	5.93D-21
NONDQUAR	1000	382	94	26963	67.04	3.76D-08
PENALTY1	1000	124	52	78	1.73	9.69D-03
POWELLSG	1000	837	281	575	6.04	5.31D-08
POWER	1000	5636	1414	1419	29.24	4.19D-09
QUARTC	1000	48	17	70	0.38	5.14D-08
SINQUAD	1000	285	119	216	9.06	1.33D-06
SROSENBR	1000	26	12	15	0.29	6.11D-20
TOINTGSS	1000	298	101	202	8.52	1.00D+01
TQUARTIC	1000	57	20	27	0.74	1.24D-13
TRIDIA	1000	16	9	576	1.40	1.31D-14
VARDIM	1000	60	15	15	0.33	2.28D-21
VAREIGVL	1000	3018	1007	1010	103.70	6.02D-11
WOODS	1000	75	24	63	0.96	6.63D-18

Key: n = number of variables, #f = number of function evaluations, #its = total number of iterations, #cg = total number of CG iterations, $time$ = total CPU time in seconds, f = smallest function value obtained.

TABLE 8
Results for the preconditioned truncated-Newton method TN(p) *on large or hard problems.*

problem	n	#f	#its	#cg	time	f
ARWHEAD	1000	20	8	10	0.68	0.00D+00
BDQRTIC	1000	30	11	15	1.17	3.98D+03
BROWNAL	100	74	26	213	7.10	1.21D-11
BRYBND	1000	72	15	41	2.65	1.91D-16
CRAGGLVY	1000	122	42	42	4.91	3.36D+02
DIXMAANA	1500	91	31	35	4.56	1.00D+00
DIXON3DQ	1000	3	1	1	0.05	0.00D+00
DQDRTIC	1000	3	1	1	0.09	0.00D+00
DQRTIC	1000	43	15	15	0.45	7.86D-08
EDENSCH	1000	42	12	13	1.30	6.00D+03
EIGENALS	110	140	36	162	2.19	8.15D-12
ENGVAL1	1000	26	10	10	0.82	1.11D+03
FLETCHCR	1000	2962	1169	1172	79.57	4.78D-14
FMINSURF	1024	248	28	1438	25.48	1.00D+00
FREUROTH	1000	47	9	1008	19.38	1.21D+05
GENROSE	1000	2290	748	1548	58.31	1.00D+00
LIARWHD	1000	48	16	27	1.24	3.51D-15
MANCINO	100	105	23	55	88.31	3.43D-21
MOREBV	1000	2	1	1	0.08	7.33D-13
NCB20B	1000	47	16	565	49.92	1.68D+03
NONDIA	1000	50	19	35	1.57	8.47D-20
NONDQUAR	1000	20	9	17	0.39	2.45D-09
PENALTY1	1000	106	32	223	9.71	9.69D-03
POWELLSG	1000	21	10	16	0.37	5.55D-08
POWER	1000	43	15	15	3.81	1.00D-10
QUARTC	1000	43	15	15	0.45	7.86D-08
SINQUAD	1000	784	307	861	37.85	2.35D-10
SROSENBR	1000	36	12	13	0.53	8.11D-25
TOINTGSS	1000	3	1	1	0.14	1.00D+01
TQUARTIC	1000	25	9	16	0.55	6.46D-21
TRIDIA	1000	3	1	1	0.05	1.97D-26
VARDIM	1000	78	28	3241	29.14	9.73D-16
VAREIGVL	1000	179	49	1932	38.61	2.35D-14
WOODS	1000	144	35	48	2.11	3.78D-15

Key: n = number of variables, $\#f$ = number of function evaluations, $\#its$ = total number of iterations, $\#cg$ = total number of CG iterations, *time* = total CPU time in seconds, f = smallest function value obtained.

TABLE 9
Results for the (unpreconditioned) limited memory method LM(n) *on large or hard problems.*

problem	n	#f	#its	#cg	time	f
ARWHEAD	1000	22	13	20	0.70	0.000D+00
BDQRTIC	1000	294	196	1104	11.84	3.984D+03
BROWNAL	100	22	9	18	0.12	2.449D-08
BRYBND	1000	31	24	80	1.39	2.152D-11
CRAGGLVY	1000	94	81	467	5.50	3.364D+02
DIXMAANA	1500	10	7	10	0.58	1.000D+00
DIXON3DQ	1000	9999	9668		343.62	5.065D-05
DQDRTIC	1000	18	11	31	0.52	9.088D-14
DQRTIC	1000	48	38	73	0.86	8.112D-06
EDENSCH	1000	31	23	77	1.29	6.003D+03
EIGENALS	110	444	408	2565	6.30	1.609D-09
ENGVAL1	1000	24	17	52	0.87	1.108D+03
FLETCHCR	1000	5795	5108	35941	255.57	7.962D-12
FMINSURF	1024	186	175	1219	11.29	1.000D+00
FREUROTH	1000	114	29	91	4.25	1.215D+05
GENROSE	1000	2460	2163	14874	111.52	1.000D+00
LIARWHD	1000	28	22	40	0.88	2.268D-16
MANCINO	100	11	9	11	15.46	8.786D-20
MOREBV	1000	75	71	526	3.32	9.964D-09
NCB20B	1000	2581	2393	18491	651.93	1.676D+03
NONDIA	1000	28	17	30	0.99	7.125D-21
NONDQUAR	1000	1073	961	7250	35.67	1.521D-05
PENALTY1	1000	121	85	166	2.76	9.687D-03
POWELLSG	1000	48	37	116	0.99	6.583D-06
POWER	1000	149	131	773	4.11	1.218D-08
QUARTC	1000	48	38	73	0.84	8.112D-06
SINQUAD	1000	141	104	385	6.41	1.129D-04
SROSENBR	1000	21	15	24	0.41	1.037D-09
TOINTGSS	1000	8	5	9	0.36	1.001D+01
TQUARTIC	1000	25	19	34	0.62	4.907D-17
TRIDIA	1000	912	871	5992	29.15	1.606D-11
VARDIM	1000	21	1	0	0.17	1.242D+22
VAREIGVL	1000	107	99	646	6.97	9.973D-09
WOODS	1000	77	52	178	2.01	1.287D-11

Key: n = number of variables, #f = number of function evaluations, #*its* = total number of iterations, #*cg* = total number of CG iterations, *time* = total CPU time in seconds, f = smallest function value obtained.

TABLE 10
Results for ISM(n,f,f) on large or hard problems, in which no preconditioning is used and the subspace is chosen from the first 10 CG directions.

problem	n	#f	#min	#its	#cg	$s^{(k)}$	time	f
ARWHEAD	1000	11	2	6	3	1.50	0.25	0.00D+00
BDQRTIC	1000	85	10	47	63	5.40	2.12	3.98D+03
BROWNAL	100	5	2	3	3	1.50	0.04	2.45D-08
BRYBND	1000	105	10	80	60	5.90	3.62	8.79D-14
CRAGGLVY	1000	108	12	90	126	7.58	4.46	3.36D+02
DIXMAANA	1500	24	7	15	7	1.00	1.09	1.00D+00
DIXON3DQ	1000	7	4	4	1748	9.00	3.97	1.25D-11
DQDRTIC	1000	10	5	5	12	2.40	0.24	1.91D-14
DQRTIC	1000	188	8	145	21	2.63	1.18	9.10D-09
EDENSCH	1000	65	7	44	25	3.57	1.76	6.00D+03
EIGENALS	110	101	8	64	86	9.00	1.16	1.90D-10
ENGVAL1	1000	52	7	34	22	3.14	1.19	1.11D+03
FLETCHCR	1000	13014	1049	10640	13187	9.92	316.24	3.60D-14
FMINSURF	1024	189	11	103	2309	10.00	21.61	1.00D+00
FREUROTH	1000	261	11	69	47	4.09	4.92	1.21D+05
GENROSE	1000	5985	411	4327	6827	9.90	138.24	1.00D+00
LIARWHD	1000	29	3	24	4	1.33	0.57	1.46D-14
MANCINO	100	68	12	37	17	1.42	67.72	1.34D-18
MOREBV	1000	4	3	3	519	10.00	1.98	1.37D-09
NCB20B	1000	174	17	101	1034	9.88	99.47	1.68D+03
NONDIA	1000	14	3	10	4	1.33	0.38	1.48D-15
NONDQUAR	1000	466	40	368	2643	9.05	10.84	2.11D-06
PENALTY1	1000	72	3	58	4	1.33	0.85	9.69D-03
POWELLSG	1000	41	3	37	9	3.00	0.35	4.92D-09
POWER	1000	105	10	80	150	7.20	1.18	1.43D-10
QUARTC	1000	188	8	145	21	2.63	1.18	9.10D-09
SINQUAD	1000	171	11	129	25	2.27	4.48	9.92D-07
SROSENBR	1000	16	2	12	3	1.50	0.17	2.45D-16
TOINTGSS	1000	7	1	4	1	1.00	0.22	1.00D+01
TQUARTIC	1000	27	2	19	3	1.50	0.34	4.05D-15
TRIDIA	1000	17	10	10	643	9.80	1.63	7.32D-15
VARDIM	1000	56	3	48	3	1.00	0.41	3.77D-23
VAREIGVL	1000	66	9	57	179	6.89	4.22	7.52D-11
WOODS	1000	75	5	42	14	2.80	0.83	9.56D-15

Key: n = number of variables, $\#f$ = number of function evaluations, $\#min$ = number of minimizations, $\#its$ = total number of iterations, $\#cg$ = total number of CG iterations, $s^{(k)}$ = average subspace dimension, $time$ = total CPU time in seconds, f = smallest function value obtained.

TABLE 11

Results for ISM(p,f,f) on large or hard problems, in which preconditioning is used and the subspace is chosen from the first 10 CG directions.

problem	n	#f	#min	#its	#cg	$s^{(k)}$	time	f
ARWHEAD	1000	23	4	17	5	1.25	0.67	0.00D+00
BDQRTIC	1000	62	7	40	7	1.00	1.55	3.98D+03
BROWNAL	100	125	10	108	134	7.70	3.31	1.05D-11
BRYBND	1000	159	11	97	26	2.36	4.58	7.78D-14
CRAGGLVY	1000	93	9	48	9	1.00	2.76	3.36D+02
DIXMAANA	1500	21	1	18	3	3.00	0.97	1.00D+00
DIXON3DQ	1000	3	1	1	1	1.00	0.05	0.00D+00
DQDRTIC	1000	3	1	1	1	1.00	0.09	1.24D-21
DQRTIC	1000	46	2	40	2	1.00	0.33	5.20D-08
EDENSCH	1000	73	6	37	6	1.00	1.73	6.00D+03
EIGENALS	110	213	13	118	58	4.46	2.24	1.25D-11
ENGVAL1	1000	32	5	21	5	1.00	0.83	1.11D+03
FLETCHCR	1000	12749	1212	10320	1212	1.00	266.31	4.00D-19
FMINSURF	1024	248	9	89	459	9.44	11.15	1.00D+00
FREUROTH	1000	113	6	40	15	2.50	2.55	1.21D+05
GENROSE	1000	6586	598	4955	985	1.65	135.03	1.00D+00
LIARWHD	1000	30	2	24	4	2.00	0.63	1.92D-14
MANCINO	100	111	13	46	14	1.08	97.22	3.03D-18
MOREBV	1000	2	1	1	1	1.00	0.08	7.33D-13
NCB20B	1000	110	15	76	656	6.67	69.90	1.68D+03
NONDIA	1000	116	7	84	17	2.43	2.74	9.90D-01
NONDQUAR	1000	86	11	73	18	1.64	0.98	2.18D-09
PENALTY1	1000	145	7	114	64	5.00	3.76	9.69D-03
POWELLSG	1000	92	14	71	16	1.14	0.94	6.14D-11
POWER	1000	61	4	51	4	1.00	1.37	3.44D-10
QUARTC	1000	46	2	40	2	1.00	0.33	5.20D-08
SINQUAD	1000	184	13	147	41	3.15	5.71	2.93D-10
SROSENBR	1000	36	7	26	7	1.00	0.50	6.14D-25
TOINTGSS	1000	3	1	1	1	1.00	0.14	1.00D+01
TQUARTIC	1000	21	2	18	3	1.50	0.36	2.01D-19
TRIDIA	1000	3	1	1	1	1.00	0.05	1.01D-25
VARDIM	1000	91	14	77	2660	6.86	20.02	5.72D-12
VAREIGVL	1000	70	9	56	154	7.33	6.98	1.25D-09
WOODS	1000	486	35	321	45	1.29	6.27	6.79D-16

Key: n = number of variables, $\#f$ = number of function evaluations, $\#min$ = number of minimizations, $\#its$ = total number of iterations, $\#cg$ = total number of CG iterations, $s^{(k)}$ = average subspace dimension, $time$ = total CPU time in seconds, f = smallest function value obtained.

TABLE 12

Results for ISM(n,e,f) *on large or hard problems, in which no preconditioning is used and the subspace is chosen from 10 extreme CG directions.*

problem	n	#f	#min	#its	#cg	$s^{(k)}$	time	f
ARWHEAD	1000	11	2	6	3	1.50	0.25	0.00D+00
BDQRTIC	1000	63	10	46	56	5.20	1.89	3.98D+03
BROWNAL	100	5	2	3	3	1.50	0.04	2.45D-08
BRYBND	1000	106	9	81	58	6.33	3.57	1.01D-12
CRAGGLVY	1000	109	12	91	113	7.58	4.42	3.36D+02
DIXMAANA	1500	24	7	15	7	1.00	1.08	1.00D+00
DIXON3DQ	1000	7	4	4	1748	9.00	4.03	1.25D-11
DQDRTIC	1000	10	5	5	12	2.40	0.24	1.91D-14
DQRTIC	1000	188	8	145	21	2.63	1.18	9.10D-09
EDENSCH	1000	65	7	44	25	3.57	1.76	6.00D+03
EIGENALS	110	108	10	70	102	8.40	1.28	1.08D-10
ENGVAL1	1000	52	7	34	22	3.14	1.19	1.11D+03
FLETCHCR	1000	12762	1075	10268	13810	9.93	314.35	8.11D-14
FMINSURF	1024	226	12	107	2140	10.00	20.82	1.00D+00
FREUROTH	1000	223	9	68	45	4.78	4.44	1.21D+05
GENROSE	1000	6642	431	4689	7088	9.90	150.00	1.00D+00
LIARWHD	1000	29	3	24	4	1.33	0.58	1.46D-14
MANCINO	100	68	12	37	17	1.42	67.98	1.34D-18
MOREBV	1000	4	3	3	519	10.00	1.99	1.37D-09
NCB20B	1000	284	18	131	1135	9.78	119.17	1.68D+03
NONDIA	1000	14	3	10	4	1.33	0.38	1.48D-15
NONDQUAR	1000	390	39	303	1636	8.62	7.79	2.67D-06
PENALTY1	1000	72	3	58	4	1.33	0.85	9.69D-03
POWELLSG	1000	41	3	37	9	3.00	0.36	4.92D-09
POWER	1000	105	10	80	150	7.20	1.20	1.43D-10
QUARTC	1000	188	8	145	21	2.63	1.19	9.10D-09
SINQUAD	1000	171	11	129	25	2.27	4.47	9.92D-07
SROSENBR	1000	16	2	12	3	1.50	0.17	2.45D-16
TOINTGSS	1000	7	1	4	1	1.00	0.22	1.00D+01
TQUARTIC	1000	27	2	19	3	1.50	0.35	4.05D-15
TRIDIA	1000	17	10	10	644	9.80	1.67	5.18D-15
VARDIM	1000	56	3	48	3	1.00	0.41	3.77D-23
VAREIGVL	1000	66	9	57	161	6.89	4.09	1.50D-10
WOODS	1000	75	5	42	14	2.80	0.83	9.56D-15

Key: n = number of variables, $\#f$ = number of function evaluations, $\#min$ = number of minimizations, $\#its$ = total number of iterations, $\#cg$ = total number of CG iterations, $s^{(k)}$ = average subspace dimension, *time* = total CPU time in seconds, f = smallest function value obtained.

TABLE 13

Results for ISM(p,e,f) on large or hard problems, in which preconditioning is used and the subspace is chosen from 10 extreme CG directions.

problem	n	#f	#min	#its	#cg	$s^{(k)}$	time	f
ARWHEAD	1000	23	4	17	5	1.25	0.67	0.00D+00
BDQRTIC	1000	62	7	40	7	1.00	1.57	3.98D+03
BROWNAL	100	122	12	107	140	7.08	3.82	6.64D-13
BRYBND	1000	159	11	97	26	2.36	4.58	7.78D-14
CRAGGLVY	1000	93	9	48	9	1.00	2.73	3.36D+02
DIXMAANA	1500	21	1	18	3	3.00	0.97	1.00D+00
DIXON3DQ	1000	3	1	1	1	1.00	0.05	0.00D+00
DQDRTIC	1000	3	1	1	1	1.00	0.09	1.24D-21
DQRTIC	1000	46	2	40	2	1.00	0.33	5.20D-08
EDENSCH	1000	73	6	37	6	1.00	1.73	6.00D+03
EIGENALS	110	213	13	118	58	4.46	2.24	1.25D-11
ENGVAL1	1000	32	5	21	5	1.00	0.82	1.11D+03
FLETCHCR	1000	12749	1212	10320	1212	1.00	265.25	4.00D-19
FMINSURF	1024	214	11	99	487	9.55	12.12	1.00D+00
FREUROTH	1000	113	6	40	15	2.50	2.55	1.21D+05
GENROSE	1000	6586	598	4955	985	1.65	135.23	1.00D+00
LIARWHD	1000	30	2	24	4	2.00	0.63	1.92D-14
MANCINO	100	111	13	46	14	1.08	97.85	3.03D-18
MOREBV	1000	2	1	1	1	1.00	0.08	7.33D-13
NCB20B	1000	117	15	78	520	6.67	62.12	1.68D+03
NONDIA	1000	116	7	84	17	2.43	2.83	9.90D-01
NONDQUAR	1000	86	11	73	18	1.64	1.00	2.18D-09
PENALTY1	1000	244	16	187	155	5.25	7.70	9.69D-03
POWELLSG	1000	92	14	71	16	1.14	0.95	6.14D-11
POWER	1000	61	4	51	4	1.00	1.38	3.44D-10
QUARTC	1000	46	2	40	2	1.00	0.34	5.20D-08
SINQUAD	1000	184	13	147	41	3.15	5.76	2.93D-10
SROSENBR	1000	36	7	26	7	1.00	0.50	6.14D-25
TOINTGSS	1000	3	1	1	1	1.00	0.14	1.00D+01
TQUARTIC	1000	21	2	18	3	1.50	0.36	2.01D-19
TRIDIA	1000	3	1	1	1	1.00	0.05	1.01D-25
VARDIM	1000	105	19	89	2842	7.21	23.54	4.61D-13
VAREIGVL	1000	84	10	61	269	7.60	8.84	3.95D-10
WOODS	1000	486	35	321	45	1.29	6.27	6.79D-16

Key: n = number of variables, $\#f$ = number of function evaluations, $\#min$ = number of minimizations, $\#its$ = total number of iterations, $\#cg$ = total number of CG iterations, $s^{(k)}$ = average subspace dimension, *time* = total CPU time in seconds, f = smallest function value obtained.

TABLE 14
Results for ISM(n,f,a) *on large or hard problems, in which no preconditioning is used and the subspace dimension is chosen automatically from the first CG directions.*

problem	n	#f	#min	#its	#cg	$s^{(k)}$	time	f
ARWHEAD	1000	11	2	6	3	2.00	0.25	0.00D+00
BDQRTIC	1000	52	11	40	62	2.18	1.69	3.98D+03
BROWNAL	100	5	2	3	3	2.00	0.04	2.45D-08
BRYBND	1000	55	10	37	74	2.40	2.27	8.21D-14
CRAGGLVY	1000	64	12	52	122	3.17	2.85	3.36D+02
DIXMAANA	1500	22	7	14	7	2.00	1.00	1.00D+00
DIXON3DQ	1000	7	4	4	1748	23.00	3.98	1.25D-11
DQDRTIC	1000	10	5	5	12	2.40	0.24	1.91D-14
DQRTIC	1000	58	9	45	18	2.00	0.38	1.75D-08
EDENSCH	1000	60	8	33	29	2.50	1.47	6.00D+03
EIGENALS	110	96	15	74	138	3.87	1.37	1.23D-11
ENGVAL1	1000	33	8	27	23	2.25	0.91	1.11D+03
FLETCHCR	1000	7720	1258	6300	14645	2.04	194.48	1.07D-12
FMINSURF	1024	190	11	106	2346	12.09	23.10	1.00D+00
FREUROTH	1000	59	9	32	25	2.44	1.75	1.21D+05
GENROSE	1000	5170	497	3674	7780	4.91	128.63	1.00D+00
LIARWHD	1000	25	5	19	8	2.00	0.54	7.19D-14
MANCINO	100	31	11	20	13	2.00	39.91	1.01D-16
MOREBV	1000	4	3	3	519	11.00	2.13	1.37D-09
NCB20B	1000	147	21	89	1095	2.71	103.56	1.68D+03
NONDIA	1000	13	3	10	4	2.00	0.37	1.48D-15
NONDQUAR	1000	280	47	231	2491	2.28	8.75	9.81D-07
PENALTY1	1000	79	14	58	19	2.00	1.03	9.69D-03
POWELLSG	1000	58	11	49	32	2.27	0.56	5.24D-11
POWER	1000	94	10	72	140	5.20	1.13	6.64D-10
QUARTC	1000	58	9	45	18	2.00	0.39	1.75D-08
SINQUAD	1000	186	34	139	67	2.21	5.91	2.02D-06
SROSENBR	1000	16	5	13	7	2.00	0.21	1.88D-15
TOINTGSS	1000	7	1	4	1	2.00	0.23	1.00D+01
TQUARTIC	1000	44	9	36	13	2.00	0.69	3.27D-16
TRIDIA	1000	17	10	10	591	9.70	1.51	7.14D-14
VARDIM	1000	58	9	40	9	2.00	0.40	1.14D-19
VAREIGVL	1000	52	9	44	162	3.78	3.66	1.77D-10
WOODS	1000	37	7	29	17	2.57	0.60	2.76D-11

Key: n = number of variables, $\#f$ = number of function evaluations, $\#min$ = number of minimizations, $\#its$ = total number of iterations, $\#cg$ = total number of CG iterations, $s^{(k)}$ = average subspace dimension, *time* = total CPU time in seconds, f = smallest function value obtained.

TABLE 15

Results for ISM(p,f,a) *on large or hard problems, in which preconditioning is used and the subspace dimension is chosen automatically from the first CG directions.*

problem	n	#f	#min	#its	#cg	$s^{(k)}$	time	f
ARWHEAD	1000	22	4	15	5	2.00	0.63	7.22D-14
BDQRTIC	1000	38	7	29	8	2.00	1.27	3.98D+03
BROWNAL	100	55	10	46	126	2.80	3.08	1.75D-11
BRYBND	1000	61	10	36	25	2.40	2.33	6.05D-14
CRAGGLVY	1000	54	10	42	10	2.00	2.06	3.36D+02
DIXMAANA	1500	20	6	15	8	2.17	1.23	1.00D+00
DIXON3DQ	1000	3	1	1	1	2.00	0.05	0.00D+00
DQDRTIC	1000	3	1	1	1	2.00	0.09	1.24D-21
DQRTIC	1000	48	8	40	8	2.00	0.41	8.23D-09
EDENSCH	1000	33	6	26	6	2.00	1.09	6.00D+03
EIGENALS	110	122	16	79	55	2.56	1.62	4.21D-11
ENGVAL1	1000	26	5	17	5	2.00	0.70	1.11D+03
FLETCHCR	1000	7053	1246	6220	1246	2.00	161.54	3.32D-16
FMINSURF	1024	199	11	77	540	5.91	11.94	1.00D+00
FREUROTH	1000	65	7	29	8	2.00	1.68	1.21D+05
GENROSE	1000	3682	593	2950	957	2.00	81.75	1.00D+00
LIARWHD	1000	124	15	65	31	2.07	2.26	1.11D+01
MANCINO	100	53	14	32	16	2.00	62.23	1.09D-18
MOREBV	1000	2	1	1	1	2.00	0.08	7.33D-13
NCB20B	1000	94	15	69	534	5.20	59.05	1.68D+03
NONDIA	1000	142	19	99	44	2.53	3.69	9.90D-01
NONDQUAR	1000	124	18	102	57	2.39	1.53	1.07D-07
PENALTY1	1000	77	13	60	60	2.08	4.30	9.69D-03
POWELLSG	1000	73	14	63	16	2.07	0.81	1.17D-09
POWER	1000	48	8	40	8	2.00	2.22	1.05D-11
QUARTC	1000	48	8	40	8	2.00	0.41	8.23D-09
SINQUAD	1000	283	56	215	177	2.29	11.50	6.90D-08
SROSENBR	1000	30	7	23	7	2.00	0.44	9.58D-26
TOINTGSS	1000	3	1	1	1	2.00	0.14	1.00D+01
TQUARTIC	1000	26	5	20	8	2.00	0.49	3.50D-16
TRIDIA	1000	3	1	1	1	2.00	0.05	1.01D-25
VARDIM	1000	78	22	57	2887	2.27	24.91	5.87D-13
VAREIGVL	1000	47	9	40	147	3.67	6.06	1.57D-10
WOODS	1000	143	23	105	25	2.00	2.19	1.21D-26

Key: n = number of variables, $\#f$ = number of function evaluations, $\#min$ = number of minimizations, $\#its$ = total number of iterations, $\#cg$ = total number of CG iterations, $s^{(k)}$ = average subspace dimension, *time* = total CPU time in seconds, f = smallest function value obtained.

TABLE 16
Results for ISM(n,e,a) on large or hard problems, in which no preconditioning is used and the subspace dimension is chosen automatically from extreme CG directions.

problem	n	#f	#min	#its	#cg	$s^{(k)}$	time	f
ARWHEAD	1000	11	2	6	3	2.00	0.25	0.00D+00
BDQRTIC	1000	52	11	40	62	2.18	1.69	3.98D+03
BROWNAL	100	5	.2	3	3	2.00	0.04	2.45D-08
BRYBND	1000	53	10	36	72	2.40	2.21	9.91D-14
CRAGGLVY	1000	66	12	55	126	3.58	3.00	3.36D+02
DIXMAANA	1500	22	7	14	7	2.00	1.00	1.00D+00
DIXON3DQ	1000	7	4	4	1748	23.00	4.19	1.25D-11
DQDRTIC	1000	10	5	5	12	2.40	0.24	1.91D-14
DQRTIC	1000	58	9	45	18	2.00	0.38	1.75D-08
EDENSCH	1000	60	8	33	29	2.50	1.47	6.00D+03
EIGENALS	110	94	17	73	127	3.47	1.35	5.34D-11
ENGVAL1	1000	33	8	27	23	2.25	0.93	1.11D+03
FLETCHCR	1000	7722	1262	6300	14691	2.03	195.36	3.98D-15
FMINSURF	1024	216	12	100	2259	8.33	22.14	1.00D+00
FREUROTH	1000	80	9	32	28	2.56	1.92	1.21D+05
GENROSE	1000	5445	475	3895	7394	5.71	126.73	1.00D+00
LIARWHD	1000	25	5	19	8	2.00	0.49	7.19D-14
MANCINO	100	31	11	20	13	2.00	38.06	1.01D-16
MOREBV	1000	4	3	3	519	11.00	1.97	1.37D-09
NCB20B	1000	159	21	95	1139	3.62	105.14	1.68D+03
NONDIA	1000	13	3	10	4	2.00	0.35	1.48D-15
NONDQUAR	1000	341	53	257	3507	2.42	11.35	1.64D-06
PENALTY1	1000	79	14	58	19	2.00	0.96	9.69D-03
POWELLSG	1000	58	11	49	32	2.27	0.52	5.24D-11
POWER	1000	94	10	72	140	5.20	1.06	6.64D-10
QUARTC	1000	58	9	45	18	2.00	0.39	1.75D-08
SINQUAD	1000	208	41	156	88	2.20	6.43	1.71D-07
SROSENBR	1000	16	5	13	7	2.00	0.19	1.88D-15
TOINTGSS	1000	7	1	4	1	2.00	0.21	1.00D+01
TQUARTIC	1000	44	9	36	13	2.00	0.65	3.27D-16
TRIDIA	1000	17	10	10	591	9.40	1.57	8.20D-14
VARDIM	1000	58	9	40	9	2.00	0.38	1.14D-19
VAREIGVL	1000	54	10	46	228	4.40	4.18	2.07D-11
WOODS	1000	37	7	29	17	2.57	0.54	2.76D-11

Key: n = number of variables, $\#f$ = number of function evaluations, $\#min$ = number of minimizations, $\#its$ = total number of iterations, $\#cg$ = total number of CG iterations, $s^{(k)}$ = average subspace dimension, $time$ = total CPU time in seconds, f = smallest function value obtained.

TABLE 17

Results for ISM(p,e,a) on large or hard problems, in which preconditioning is used and the subspace dimension is chosen automatically from extreme CG directions.

problem	n	#f	#min	#its	#cg	$s^{(k)}$	time	f
ARWHEAD	1000	22	4	15	5	2.00	0.59	7.22D-14
BDQRTIC	1000	38	7	29	8	2.00	1.17	3.98D+03
BROWNAL	100	56	11	48	140	3.27	3.28	3.25D-13
BRYBND	1000	66	11	40	26	2.27	2.58	2.06D-14
CRAGGLVY	1000	54	10	42	10	2.00	2.04	3.36D+02
DIXMAANA	1500	20	6	15	8	2.17	1.20	1.00D+00
DIXON3DQ	1000	3	1	1	1	2.00	0.05	0.00D+00
DQDRTIC	1000	3	1	1	1	2.00	0.09	1.24D-21
DQRTIC	1000	48	8	40	8	2.00	0.42	8.23D-09
EDENSCH	1000	33	6	26	6	2.00	1.08	6.00D+03
EIGENALS	110	112	16	70	56	2.44	1.53	2.50D-13
ENGVAL1	1000	26	5	17	5	2.00	0.70	1.11D+03
FLETCHCR	1000	7053	1246	6220	1246	2.00	162.80	3.32D-16
FMINSURF	1024	190	13	86	549	6.54	12.77	1.00D+00
FREUROTH	1000	65	7	29	8	2.00	1.67	1.21D+05
GENROSE	1000	3682	593	2950	957	2.00	81.89	1.00D+00
LIARWHD	1000	124	15	65	31	2.07	2.24	1.11D+01
MANCINO	100	53	14	32	16	2.00	64.59	1.09D-18
MOREBV	1000	2	1	1	1	2.00	0.08	7.33D-13
NCB20B	1000	101	15	70	505	4.00	57.74	1.68D+03
NONDIA	1000	142	19	99	44	2.53	3.66	9.90D-01
NONDQUAR	1000	122	17	96	46	2.47	1.46	9.13D-08
PENALTY1	1000	80	14	64	62	2.07	4.60	9.69D-03
POWELLSG	1000	73	14	63	16	2.07	0.80	1.17D-09
POWER	1000	48	8	40	8	2.00	2.24	1.05D-11
QUARTC	1000	48	8	40	8	2.00	0.41	8.23D-09
SINQUAD	1000	440	93	301	281	2.15	17.97	1.11D-08
SROSENBR	1000	30	7	23	7	2.00	0.44	9.58D-26
TOINTGSS	1000	3	1	1	1	2.00	0.14	1.00D+01
TQUARTIC	1000	26	5	20	8	2.00	0.49	3.50D-16
TRIDIA	1000	3	1	1	1	2.00	0.05	1.01D-25
VARDIM	1000	82	22	59	2804	2.32	24.92	1.24D-12
VAREIGVL	1000	49	9	42	134	3.67	6.03	3.26D-10
WOODS	1000	143	23	105	25	2.00	2.20	1.21D-26

Key: n = number of variables, $\#f$ = number of function evaluations, $\#min$ = number of minimizations, $\#its$ = total number of iterations, $\#cg$ = total number of CG iterations, $s^{(k)}$ = average subspace dimension, *time* = total CPU time in seconds, f = smallest function value obtained.

Chapter 6
Conjugate Directions in Perspective

William C. Davidon[*]

Abstract

With an appropriate choice for a conjugacy relation on the set of lines in the domain $X \subseteq \mathbb{R}^n$ of an objective function $f : X \to \mathbb{R}$, the search for an n-dimensional minimizer reduces to a sequence of n line searches in conjugate directions. This paper defines appropriate conjugacy relations for conic generalizations of quadratic functions, and interprets these relations geometrically.

1 Conic Functions

Conic functions [2, 4] generalize quadratics, and can better match the values and gradients of typical objective functions, particularly those like penalty functions which rise rapidly near some hyperplane.

DEFINITION 1.1. *A smooth real-valued function over an open convex subset in $\mathbb{R}^n$ is affine iff it has constant gradient, quadratic iff it has constant hessian, and conic iff it is the ratio of a quadratic to the square of an affine function.*

Function $f : X \to \mathbb{R}$ with open convex domain $X \subseteq \mathbb{R}^n$ is conic and has value $f_* \in \mathbb{R}$ and hessian $A_* \in \mathbb{R}^{n \times n}$ at a critical point $x_* \in X$ iff there is an $a_* \in \mathbb{R}^n$ with

$$f(x) = f_* + \frac{(x - x_*)^T A_*(x - x_*)}{2(1 - a_*^T(x - x_*))^2}$$

and $a_*^T(x - x_*) < 1$ for all $x \in X$. The critical point x_* is unique iff the symmetric matrix A_* is invertible, and x_* is a minimizer iff A_* is non-negative. The domain X of f is maximal — i.e., f has no smooth extension to a larger convex domain — iff $X = \{x \in \mathbb{R}^n : a^T(x - x_*) < 1\}$. We now assume X is maximal and $x_* \in X$ is a unique minimizer. Function f is quadratic iff $a_* = 0$, iff it is convex, and iff all its level sets are ellipsoids centered at x_*. If $a_* \ne 0$ then the level set $f^{-1}[f_* + \lambda/a_*^T A_*^{-1} a_*] \subset X$ is an ellipsoid iff $0 < \lambda < 1/2$, a paraboloid iff $\lambda = 1/2$, and a lobe of a hyperboloid iff $\lambda > 1/2$. While the hessian of f at x_* is the positive definite A_*, the hessian of f at $x_* + s$ is positive definite iff $a_*^T A_*^{-1} a_* s^T A_* s < (1 - a_*^T s)^2$.

For each point p in the open convex domain $X \subseteq \mathbb{R}^n$ of a conic function $f : X \to \mathbb{R}$ there is just one $f_p \in \mathbb{R}$, $g_p \in \mathbb{R}^n$, $a_p \in \mathbb{R}^n$ and symmetric $A_p \in \mathbb{R}^{n \times n}$ with

$$f(x) = f_p + \frac{g_p^T(x - p)}{1 - a_p^T(x - p)} + \frac{(x - p)^T A_p(x - p)}{2(1 - a_p^T(x - p))^2}$$

and $a_p^T(x - p) < 1$ for all $x \in X$. This function has value f_p, gradient g_p, and hessian $A_p + a_p g_p^T + g_p a_p^T$ at point p. For any points $p, q \in X$:

---
[*]Haverford College, Haverford PA 19041.

79

$$\begin{aligned}
f_q &= f_p + g_p^T z_{pq} + \tfrac{1}{2} z_{pq}^T A_p z_{pq}, \\
g_q &= \Gamma_{pq}^T (g_p + A_p z_{pq}), \\
A_q &= \Gamma_{pq}^T A_p \Gamma_{pq},
\end{aligned}$$
and $a_q = a_p / \gamma_{pq}$;
where $\gamma_{pq} = 1 - a_p^T(q-p)$, $z_{pq} = (q-p)/\gamma_{pq}$, and $\Gamma_{pq} = (I + z_{pq} a_p^T)/\gamma_{pq}$. For any points $p, q, r \in X$, $\gamma_{pq} \gamma_{qr} = \gamma_{pr}$ and $\Gamma_{pq} \Gamma_{qr} = \Gamma_{pr}$.

DEFINITION 1.2. *The line through point $p \in X$ in direction $u \in \mathbb{R}^n$ and the line through point $q \in X$ in direction $v \in \mathbb{R}^n$ are conjugate with respect to the conic function*

$$f : X \ni x \mapsto f_p + \frac{g_p^T(x-p)}{1 - a_p^T(x-p)} + \frac{(x-p)^T A_p (x-p)}{2(1 - a_p^T(x-p))^2} \in \mathbb{R}$$

iff $u^T A_p (I - (q-p)a_p^T)^{-1} v = 0$; A_p is the conjugacy matrix *for f at p.*
Note that $(I - (q-p)a_p^T)^{-1} = I + (q-p)a_p^T/(1 - a_p^T(q-p)) = I - (p-q)a_q^T$.

A conic function $f : X \to \mathbb{R}$ has unique minimizer $x_* \in X$ iff each sequence of n exact line searches along conjugate lines in X terminates at x_*. While the conjugacy matrix A_* and hessian at the minimizer $x_* \in X$ of conic $f : X \to \mathbb{R}$ are equal, they are equal at other points in X iff f is quadratic. For nonquadratic conic $f : X \to \mathbb{R}$, the Newton step from almost all points of X is not in the direction of the minimizer x_*. It is the conjugacy matrix A_p at each point $p \neq x_*$, and not the hessian, which determines the direction from p to x_*, since for all $p \in X$, $A_p(p - x_*)$ is a positive multiple of the gradient at p.

2 Collinear Scalings

DEFINITION 2.1. *A bijection $S : W \to X$ between open convex subsets in $\mathbb{R}^n$* scales *$f : X \to \mathbb{R}$ iff the composition $f \circ S : W \to \mathbb{R}$ is quadratic with unit hessian. Scaling $S : W \to X$ is* collinear *iff each line in W has a linear image in X, and* affine *iff its jacobian is constant.*

Morse's Lemma [3] says that if a smooth function f has a positive definite hessian at a local minimizer x_*, then some restriction of f to a neighborhood of x_* has a smooth scaling. Functions with unique minimizers have affine scalings iff they are quadratic, and they have collinear scalings iff they are conic.

A bijection with open convex domain $W \subseteq \mathbb{R}^n$ and range $X \subseteq \mathbb{R}^n$ is collinear with value $x_* \in X$ and jacobian $J \in \mathbb{R}^{n \times n}$ at $p \in W$ iff it equals

$$W \ni w \mapsto x_* + \frac{J(w-p)}{1 + h^T(w-p)} \in X$$

for some $h \in \mathbb{R}^n$ with $h^T(w-p) > -1$ for all $w \in W$. This bijection scales the conic function

$$X \ni x \mapsto f_* + \frac{(x - x_*)^T A_*(x - x_*)}{1 - a_*^T(x - x_*)^2} \in \mathbb{R}$$

iff $h = J^T a_*$ and $J^T A_* J = I$.

Each collinear scaling $S : W \to X$ of a conic function $f : X \to \mathbb{R}$ with minimizer $x_* \in X$ pairs concentric n-dimensional spheres in W with the level sets of f, it pairs the common center of these spheres with x_*, and it pairs orthogonal lines in W with lines in X which are conjugate with respect to f.

All affine bijections are collinear, and a collinear bijection is affine iff it has a collinear extension to all of $\mathbb{R}^n$. A collinear permutation of the interior of an n-dimensional ellipsoid is affine iff it leaves the center fixed. Affine permutations of this interior form an $n(n-1)/2$ dimensional group isomorphic to the orthogonal group $O(n)$. Collinear permutations of this interior form an $n(n+1)/2$ dimensional group isomorphic to the group $O(n,1)$ of isometries of n-dimensional hyperbolic space. This group acts transitively both on the interior of the ellipsoid and on its set of secants; *i.e.*, all points inside the ellipsoid can be obtained from any one by collinear permutations, and all secants can also be so obtained from any secant.

For example, a permutation of $O_n = \{x \in \mathbf{R}^n : x^T x < 1\}$ is collinear and has value $p \in O_n$ at $0 \in O_n$ iff it is the product of an isometry of the sphere followed by the permutation

$$O_n \ni x \mapsto p + \frac{\sqrt{1-p^T p}}{1+p^T x}(x - \frac{pp^T x}{1+\sqrt{1-p^T p}}) \in O_n.$$

3 Geometry of Conjugacy

Conjugacy of lines with respect to a conic function is related to the concept from projective geometry of conjugacy of lines with respect to a conic section [1]. It suffices here to consider only the conjugacy of secants in ellipsoids.

DEFINITION 3.1. *Two secants in an ellipsoid are* conjugate *iff two tangents with distinct points of tangency on one secant either meet at a point on the other secant or else are both parallel to other secant. Secants in an ellipsoid are* x_**-conjugate iff they are images of orthogonal lines under a collinear bijection from a sphere to the ellipsoid which pairs the center of the sphere with point* x_*.

Lines through the minimizer of a conic function f are conjugate with respect to f iff they are conjugate with respect to any of the ellipsoidal level sets of f, and iff they are x_*-conjugate. Each point x_* inside an ellipsoid determines a symmetric and irreflexive x_*-conjugacy relation on the set of all secants in the ellipsoid. No two points in an ellipsoid determine the same conjugacy relation. While not all ellipsoids about a given point x_* determine the same conjugacy relation, we leave implicit the dependence of the x_* conjugacy relation on the ellipsoid.

If point x_* is the center of an n-dimensional sphere then x_*-conjugacy is just orthogonality. If three secants in an ellipsoid are the sides of a triangle and one goes through point x_* then the opposite two sides are x_*-conjugate. If x_* is any point inside an ellipsoid then x_*-conjugacy is conjugacy with respect to any of the two dimensional family of conic functions with minimizer x_* for which the ellipsoid is a level set. These conic functions are quadratic iff x_* is the center of the ellipsoid.

For each pair of points p and x_* inside an ellipsoid there is a positive definite and symmetric matrix $A_p \in \mathbf{R}^{n \times n}$ for which lines through p in directions $u, v \in \mathbf{R}^n$ are x_*-conjugate iff $u^T A_p v = 0$. Conversely, for each pair of points $p, x_* \in \mathbf{R}^n$ and positive definite matrix $A_p \in \mathbf{R}^{n \times n}$ there are ellipsoids for which lines through p in directions $u, v \in \mathbf{R}^n$ are x_*-conjugate iff $u^T A_p v = 0$. Each of these ellipsoids is a level set of a conic function with minimizer x_* and conjugacy matrix A_p at p, and each of these conic functions equals

$$x \mapsto f_p - \frac{(x_*-p)^T A_p(x-p)}{(1-a_p^T(x_*-p))(1-a_p^T(x-p))} + \frac{(x-p)^T A_p(x-p)}{2(1-a_p^T(x-p))^2}$$

for some $f_p \in \mathbf{R}$ and $a_p \in \mathbf{R}^n$ with $a_p^T(x_*-p) < 1$.

References

[1] H. S. M. Coxeter, *Projective Geometry*, Blaisdell Publishing Company, 1964.
[2] W. C. Davidon, *Conic approximations and collinear scalings for optimizers*, Siam J. Numer. Anal. 17 (1980), pp. 268–281.
[3] M. Morse *The calculus of variation in the large*, American Mathematical Society, Providence RI, 1934.
[4] J. L. Nazareth and K. A. Ariyawansa, *On accelerating Newton's method based on a conic model*, Information Processing Letters 30 (1989), pp. 277–281.

Chapter 7
Krylov Subspace Approximations to the Solution of a Linear System

Anne Greenbaum[*]

Abstract

This paper deals with the problems of finding the optimal approximation to the solution of a linear system from a Krylov subspace and giving an estimate of how good that approximation will be, based on simple properties of the coefficient matrix. Some recent advances and interesting open problems in this field are discussed. Topics addressed include the behavior of the conjugate gradient algorithm in finite precision arithmetic and *a priori* bounds on the residual norm in the GMRES method.

1 Introduction

Given a nonsingular n by n linear system $Ax = b$ and an initial guess x_0 for the solution many iterative methods generate successive approximations $x_1, x_2, \ldots$ satisfying

$$(1) \qquad x_k \in x_0 + \text{span}\{r_0, Ar_0, \ldots, A^{k-1}r_0\}, \quad k = 1, 2, \ldots$$

where $r_0 = b - Ax_0$ is the initial residual. The space $\text{span}\{r_0, Ar_0, \ldots A^{k-1}r_0\}$ is referred to as a *Krylov subspace*, generated by A and r_0. Given that x_k is to be of the form (1), one must ask the following two questions:

1. How good an approximate solution is contained in the affine space (1)?

2. How can a good (optimal) approximation from this space be computed with a moderate amount of work and storage?

These questions are the subject of this paper.

If it turns out that the space (1) does not contain a good approximate solution for any moderate size k, or if such an approximate solution cannot be easily computed, then one might consider modifying the original problem to obtain a better Krylov subspace. For example, one might use a *preconditioner* M and effectively solve the modified problem

$$M^{-1}Ax = M^{-1}b$$

by generating approximate solutions $x_1, x_2,, \ldots$ satisfying

$$x_k \in x_0 + \text{span}\{M^{-1}r_0, (M^{-1}A)M^{-1}r_0, \ldots, (M^{-1}A)^{k-1}M^{-1}r_0\}, \quad k = 1, 2, \ldots.$$

At each step of the preconditioned algorithm, it is necessary to compute the product of M^{-1} with a vector, or, equivalently, to solve a linear system with coefficient matrix M,

---

[*]Courant Institute of Mathematical Sciences, 251 Mercer St., New York, NY 10012. This work was supported by DOE contract DEFG0288ER25053.

so M should be chosen so that such linear systems are much easier to solve than the original problem. The subject of finding good preconditioners is a large one on which much interesting research has focused in recent years. We will not attempt to address this topic here, however, simply because it is so broad. Instead, we address the question of what properties one should look for in a preconditioned matrix, to ensure the existence of a good approximate solution in a relatively low dimensional Krylov space.

1.1 Hermitian Matrices

For *real symmetric* or *complex Hermitian* matrices A, the answers to questions 1 and 2 are well-understood — at least in exact arithmetic. There is a known short recurrence for finding the 'optimal' approximation of the form (1), if 'optimal' is taken to mean the approximation whose residual, $b - Ax_k$, has the smallest Euclidean norm. An algorithm that generates this optimal approximation is called MINRES (minimal residual) [21]. If A is also positive definite, then one might instead minimize the A-norm of the error, $\|e_k\|_A \equiv \langle A^{-1}b - x_k, b - Ax_k \rangle^{1/2}$. The conjugate gradient (CG) algorithm [15] generates this approximation. For each of these algorithms, the work required at each iteration is the computation of one matrix-vector product (which is always required to generate the next vector in the Krylov space) plus a few vector inner products, and only a few vectors need be stored. Since these methods find the 'optimal' approximation with little extra work and storage, beyond that required for the matrix-vector multiplication, they are almost always the methods of choice for Hermitian problems.

Additionally, we can describe precisely how good these optimal approximations are (for the worst possible initial vector r_0), in terms of the *eigenvalues* of the matrix. Consider the 2-norm of the residual in the MINRES algorithm. It follows from (1) that the residual r_k can be written in the form

$$r_k = P_k(A)r_0,$$

where P_k is a certain k^{th} degree polynomial with value 1 at the origin, and, for any other such polynomial p_k, we have

(2) $$\|r_k\| \leq \|p_k(A)r_0\|.$$

Writing the eigendecomposition of A as $A = Q\Lambda Q^*$, where $\Lambda = \text{diag}(\lambda_1, \ldots, \lambda_n)$ is a diagonal matrix of eigenvalues and Q is a unitary matrix whose columns are eigenvectors, expression (2) implies

$$\|r_k\| \leq \|Qp_k(\Lambda)Q^* r_0\| \leq \|p_k(\Lambda)\| \, \|r_0\|,$$

and since this holds for any k^{th} degree polynomial p_k with $p_k(0) = 1$, we have

(3) $$\|r_k\|/\|r_0\| \leq \min_{p_k} \max_{i=1,\ldots,n} |p_k(\lambda_i)|.$$

It turns out that the bound (3) on the size of the MINRES residual at step k is *sharp* — that is, for each k there is a vector r_0 for which this bound will be attained [9, 11, 16]. Thus the question of the size of the MINRES residual at step k is reduced to a problem in approximation theory — how well can one approximate zero on the set of eigenvalues of A using a kth degree polynomial with value 1 at the origin? One can answer this question precisely with a complicated expression involving all of the eigenvalues of A or one can give simple bounds in terms of just a few of the eigenvalues of A. The important point is that the norm of the residual (for the worst-case initial vector) is completely determined by the eigenvalues of the matrix, and we have at least intuitive ideas of what constitute good and bad eigenvalue distributions. The same reasoning shows that the A-norm of the error in the CG algorithm satisfies

$$\|e_k\|_A/\|e_0\|_A \leq \min_{p_k} \max_{i=1,\ldots,n} |p_k(\lambda_i)|, \tag{4}$$

and, again, this bound is sharp.

Thus, for Hermitian problems, we not only have good algorithms for finding the optimal approximation from the Krylov space (1) but we can also say just how good that approximation will be, based on simple properties of the coefficient matrix (i.e., the eigenvalues). It is therefore fair to say that the iterative solution of Hermitian linear systems is well-understood — except for one thing. All of the above discussion assumes *exact arithmetic*. It is well-known that in finite precision arithmetic the CG and MINRES algorithms do *not* find the optimal approximation from the Krylov space (1) or necessarily anything close to it! In fact, the CG algorithm originally lost favor partly because it did not behave the way exact arithmetic theory predicted [4]. In section 2 we discuss the behavior of the CG algorithm for Hermitian positive definite problems and the related Lanczos algorithm [17] for general Hermitian problems in finite precision arithmetic.

1.2 Nonhermitian Matrices

While the methods of choice and convergence analysis for Hermitian problems are well-understood in exact arithmetic, and the results of section 2 go a long way towards explaining the behavior of these methods in finite precision arithmetic, the state-of-the-art for *nonhermitian* problems is not nearly so well-developed. One difficulty is that no method is known for finding the optimal approximation from the space (1), while performing only $O(n)$ operations per iteration, in addition to the matrix-vector multiplication. In fact, a theorem due to Faber and Manteuffel [6] shows that for most nonhermitian matrices A there is no short recurrence for the optimal approximation from the Krylov space (1). To fully understand this result, one must consider the statement and hypotheses carefully, and we give more of these details in section 4. Still, the current options for nonhermitian problems are either to perform extra work ($O(nk)$ operations at step k) and use extra storage ($O(nk)$ words to perform k iterations) to find the optimal approximation from the Krylov space or to settle for a nonoptimal approximation. The (full) GMRES (generalized minimal residual) algorithm [22] (and other mathematically equivalent algorithms) find the approximation of the form (1) for which the 2-norm of the residual is minimal, at the cost of this extra work and storage, while other nonhermitian iterative methods (e.g., BCG [7], CGS [23], QMR [8], BiCGSTAB [27], restarted GMRES [22], hybrid GMRES [18], etc.) generate nonoptimal approximations.

Even if one does generate the optimal approximation of the form (1), we are still unable to answer question 1. That is, there is no known way to describe how good this optimal approximation will be (for the worst-case initial vector), in terms of simple properties of the coefficient matrix. One might try an approach based on eigenvalues, as was done for the Hermitian case. Assume that A is diagonalizable and write an eigendecomposition of A as $A = V\Lambda V^{-1}$, where Λ is a diagonal matrix of eigenvalues and the columns of V are eigenvectors. Then it follows from (2) that the GMRES residual at step k satisfies

$$\text{(5)} \qquad \|r_k\| \leq \min_{p_k} \|V p_k(\Lambda) V^{-1} r_0\| \leq \kappa(V) \min_{p_k} \max_{i=1,\ldots,n} |p_k(\lambda_i)| \, \|r_0\|,$$

where $\kappa(V) = \|V\| \cdot \|V^{-1}\|$ is the condition number of the eigenvector matrix. The scaling of the columns of V can be chosen to minimize this condition number. When A is a *normal* matrix, so that V can be taken to have condition number one, it turns out that the bound (5) is sharp, just as for the Hermitian case. Thus for normal matrices, the analysis of GMRES again reduces to a question of polynomial approximation — how well can one approximate zero on the set of (complex) eigenvalues using a kth degree polynomial with value 1 at the origin. When A is nonnormal, however, the bound (5) may *not* be sharp. If the condition number of V is huge, the right-hand side of (5) may also be large, but this does not necessarily imply that GMRES converges poorly. It may simply mean that the bound (5) is a large overestimate of the actual GMRES residual norm. An interesting open question is to determine when an ill-conditioned eigenvector matrix implies poor convergence for GMRES and when it simply means that the bound (5) is a large overestimate. If eigenvalues are not the key, then one would like to be able to describe the behavior of GMRES in terms of some other characteristic properties of the matrix A. Some ideas along these lines are discussed in section 3.

Finally, since the full GMRES algorithm may be impractical if a fairly large number of iterations are required, one would like to have theorems relating the convergence of some nonoptimal methods (that do not require extra work and storage) to that of GMRES. Unfortunately, no fully satisfactory theorems of this kind are currently available, and this important open problem is discussed in section 4.

2 The Lanczos and Conjugate Gradient Algorithms in Finite Precision Arithmetic

In [13], a model was suggested for understanding the behavior of the conjugate gradient algorithm in finite precision arithmetic. Consider the exact CG algorithm applied to a linear operator $\mathcal{A}$ with infinitely many eigenvalues, but whose eigenvalues all lie in tiny intervals about the eigenvalues of A, the size of the intervals being a function of the machine precision. Error bounds that apply to the exact algorithm for solving the operator equation $\mathcal{A}u = \varphi$ can be expected to hold for the finite precision computation for $Ax = b$, where $\|\mathcal{A}^{-1}\varphi - u_0\|_{\mathcal{A}} \approx \|A^{-1}b - x_0\|_A$. Thus, the polynomial approximation bound (4) is replaced by

$$\text{(6)} \qquad \|e_k\|_A / \|e_0\|_A \leq \min_{p_k} \max_{z \in \bigcup_{i=1}^n [\lambda_i - \delta, \lambda_i + \delta]} |p_k(z)|,$$

where δ is the width of the intervals containing the eigenvalues of $\mathcal{A}$. Sometimes the right-hand side of (6) is approximately the same as that of (4) and sometimes it is quite different. These two bounds are plotted in Fig. 1, for a problem with eigenvalues

$$\lambda_1 = 10^{-3}, \quad \lambda_n = 1, \quad \lambda_i = \lambda_1 + \frac{i-1}{n-1}(\lambda_n - \lambda_1)\rho^{n-i}, \quad i = 2, \ldots, n-1,$$

where $n = 24$ and $\rho = .6$. The eigenvalues of this matrix are tightly clustered at the lower end of the spectrum and well spread out at the upper end. The interval width δ was taken to be 2×10^{-15}, or, approximately $20\epsilon \|A\|$, where ϵ is the machine precision. For such problems, the bounds in (4) and (6) tend to differ substantially, as can be seen in the figure. Also shown is the A-norm of the error at each step of the conjugate gradient

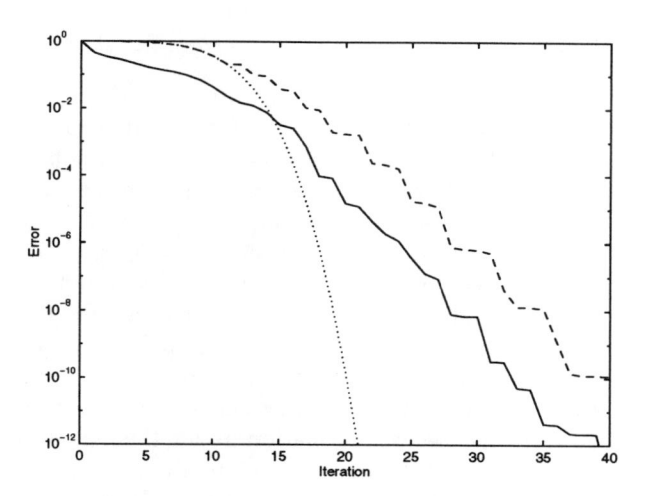

FIG. 1. *Exact Arithmetic Error Bound (dotted), Finite Precision Arithmetic Error Bound (dashed), and Actual Error in a Finite Precision Computation (solid)*

algorithm applied to a linear system $Ax = b$, with eigenvalues $\lambda_1, \ldots, \lambda_n$, using MATLAB on a Sparc workstation, with machine precision $\epsilon \approx 10^{-16}$. Note that the finite precision computation satisfies the bound in (6) but not the one in (4) and that (6) provides a fairly good estimate of the actual error in the finite precision computation.

An estimate of the form (6) has been proved rigorously [10], but the proven result is not as nice as one might hope. The proven bound on the interval size δ is a large overestimate and, additionally, it depends on k. Thus, the model described above has been proved to hold for a fixed value δ only when the number of steps k does not exceed some upper bound. Recent work has focused on trying to establish error bounds for finite precision computations that give useful information for larger values of k or even for infinitely many steps k [3, 2].

In [2], an analysis is developed assuming that the main effect of roundoff is on the underlying three-term Lanczos recurrence. That is, in exact arithmetic, the approximate solution x_k produced by the conjugate gradient algorithm can be written in the form $x_k = x_0 + Q_k y$, where y is the solution of a k by k tridiagonal system, $T_k y = \beta u_1$, where $\beta = \|r_0\|$ and $u_1 = (1, 0, \ldots, 0)^T$, and the columns of Q_k form an orthonormal basis for the Krylov space in (1). If Q_k and T_k are generated by a finite precision Lanczos computation, then the columns of Q_k may not be orthogonal. One might still ask how accurate is the approximate solution $x_k \equiv x_0 + Q_k y$, when Q_k is produced by a finite precision Lanczos computation and when y satisfies $T_k y = \beta u_1$, where T_k is the tridiagonal matrix produced by the finite precision Lanczos computation. This idea of separating the effect of rounding errors on the Lanczos process from that of other aspects of the conjugate gradient algorithm was suggested, for example, in [20].

For this hypothetical implementation, it can be shown that the 2-norm of the residual r_k is almost completely determined by the tridiagonal matrix T_k and the next coefficient β_k produced in the finite precision Lanczos computation, at least at steps k where $\|T_k^{-1}\|$ is not too large. More specifically, $\|r_k\|/\|r_0\|$ is approximately equal to the absolute value of β_k times the $(k, 1)$-entry of T_k^{-1}:

$$\|r_k\|/\|r_0\| \approx |\beta_k(u_k^T T_k^{-1} u_1)|.$$

Suppose the exact Lanczos algorithm, applied to a matrix $\bar{A}$ with initial vector $\bar{q}_1$, generates the same tridiagonal matrix T_k and the same coefficient β_k as the finite precision computation for A with initial vector q_1. It would follow that the 2-norm of the residual r_k in the finite precision computation would be approximately the same as the 2-norm of the residual $\bar{r}_k$ in the exact algorithm for solving a linear system $\bar{A}\bar{x} = \bar{b}$, with initial residual $\bar{r}_0 = \|r_0\|\bar{q}_1$. Note also that if T is any Hermitian tridiagonal matrix whose k by k principal submatrix is T_k and whose $(k, k+1)$ and $(k+1, k)$ entries are β_k, then the exact Lanczos algorithm applied to T with initial vector u_1 will generate T_k and β_k at step k. With this observation, one can now use results about the convergence rate of the exact Lanczos algorithm applied to any such matrix T to derive bounds on the residual r_k in the finite precision computation. To do this, one needs information about the eigenvalues of such matrices T. Information of this sort is derived and used in [2] to give bounds on the 2-norm of the residual in finite precision computations with both positive definite and Hermitian indefinite matrices A.

3 Convergence Rate of GMRES

As noted in the introduction, the convergence behavior of the GMRES method is *not* determined by the eigenvalues of the matrix, for most nonnormal problems. In fact, it is shown in [14, 12] that *any* nonincreasing curve represents a plot of residual norm versus iteration number for the GMRES method applied to some problem, and, moreover, that problem can be taken to have any desired eigenvalues. Thus, for example, eigenvalues tightly clustered around 1 are not necessarily good for nonnormal matrices, as they are for normal ones.

A simple way to see this is to consider a matrix A with the following sparsity pattern:

(7)
$$\begin{pmatrix} 0 & * & 0 & \cdots & 0 \\ 0 & * & * & \cdots & 0 \\ \vdots & \vdots & \vdots & \ddots & \\ 0 & * & * & \cdots & * \\ * & * & * & \cdots & * \end{pmatrix}$$

If the initial residual r_0 is a multiple of the first unit vector $u_1 \equiv (1, 0, \ldots, 0)^T$, then Ar_0 is a multiple of u_n, $A^2 r_0$ is a linear combination of u_n and u_{n-1}, etc. All vectors $A^k r_0$, $k = 1, \ldots, n-1$ are orthogonal to r_0, so the optimal approximation of the form (1) is simply $x_k = x_0$, for $k = 1, \ldots, n-1$; i.e., GMRES makes no progress until step n! Now, the class of matrices of the form (7) includes, for example, all *companion matrices*:

$$\begin{pmatrix} 0 & 1 & & & \\ 0 & 0 & \ddots & & \\ \vdots & \vdots & & 1 & \\ c_0 & c_1 & \cdots & c_{n-1} \end{pmatrix}$$

The eigenvalues of this matrix are the roots of the polynomial $z^n - \sum_{j=0}^{n-1} c_j z^j$, and the coefficients $c_0, \ldots, c_{n-1}$ can be chosen to make this matrix have any desired eigenvalues.

Of course, one would probably not use the GMRES algorithm to solve a linear system with the sparsity pattern of that in (7), but the same result holds for any matrix that is unitarily similar to one of the form (7). Note that (7) is simply a permuted upper triangular matrix. Every matrix is unitarily similar to an upper triangular matrix, but fortunately most matrices are *not* unitarily similar to one of the form (7)!

If eigenvalues are not the answer, then one must consider other characteristic properties of a matrix in order to describe the behavior of GMRES. Trefethen has suggested pseudo-eigenvalues [26]. The argument is as follows. Any function $p(A)$ of a matrix A can be written as a Cauchy integral

$$(8) \qquad p(A) = \frac{1}{2\pi i} \int_\Gamma p(z)(zI - A)^{-1} dz,$$

where Γ is any simple closed curve or union of simple closed curves containing the spectrum of A. Taking norms on each side in (8) and replacing the norm of the integral by the length of the curve times the maximum norm of the integrand gives

$$(9) \qquad \|p(A)\| \leq \frac{\mathcal{L}(\Gamma)}{2\pi} \max_{z \in \Gamma} \|p(z)(zI - A)^{-1}\|.$$

Now, if we consider a curve Γ_ϵ on which the resolvent norm $\|(zI - A)^{-1}\|$ is constant, say, $\|(zI - A)^{-1}\| = \epsilon^{-1}$, then (9) implies

$$(10) \qquad \|p(A)\| \leq \frac{\mathcal{L}(\Gamma_\epsilon)}{2\pi\epsilon} \max_{z \in \Gamma_\epsilon} |p(z)|.$$

The curve on which $\|(zI - A)^{-1}\| = \epsilon^{-1}$ is referred to as the boundary of the ϵ-pseudospectrum of A:

$$\Lambda_\epsilon \equiv \{z : \|(zI - A)^{-1}\| \geq \epsilon^{-1}\}.$$

From (10), and the optimality of the GMRES approximation x_k in (1), it follows that the GMRES residual r_k satisfies

$$(11) \qquad \|r_k\|/\|r_0\| \leq \frac{\mathcal{L}(\Gamma_\epsilon)}{2\pi\epsilon} \min_{p_k} \max_{z \in \Gamma_\epsilon} |p_k(z)|,$$

for any choice of the parameter ϵ. For certain problems, and with carefully chosen values of ϵ, the bound (11) may be much smaller than that in (5). Still, the bound (11) is not sharp, and for some problems there is no choice of ϵ that yields a realistic estimate of the actual GMRES residual [14]. It is easy to see where the main overestimate occurs. In going from (8) to (9), replacing the norm of the integral by the length of the curve times the maximum norm of the integrand, one may lose important cancellation properties of the integral.

Both of the inequalities (5) and (11) provide bounds on the GMRES residual by bounding the quantity $\min_{p_k} \|p_k(A)\|$. Now, the worst-case behavior of GMRES is given by

$$(12) \qquad \|r_k\| = \max_{\|r_0\|=1} \min_{p_k} \|p_k(A)r_0\|.$$

The polynomial p_k depends on r_0. Until recently, it was an open question whether the right-hand side of (12) was equal to the quantity

$$(13) \qquad \min_{p_k} \|p_k(A)\| \equiv \min_{p_k} \max_{\|r_0\|=1} \|p_k(A)r_0\|.$$

It is known that the right-hand sides of (12) and (13) are equal if A is a normal matrix or if the dimension of A is less than or equal to 3 or if $k = 1$, and many numerical experiments have shown that these two quantities are equal (to within the accuracy limits of the computation) for a wide variety of matrices and all values of k. Recently, however, it has been shown that the two quantities may differ. Faber, *et. al.* [5] constructed an example in which the right-hand side of (13) is 1, while that of (12) is .9995. Subsequently, Toh [24] generated examples in which the ratio of the right-hand side of (13) to that of (12) can be made arbitrarily large by varying a parameter in the matrix. Thus neither of the approaches leading to inequalities (5) and (11) can be expected to yield a sharp bound on the size of the GMRES residual, and it remains an open problem to describe the convergence of GMRES in terms of some simple characteristic properties of the coefficient matrix.

4 Nonoptimal Krylov Subspace Approximations

Because GMRES requires extra work and storage as the iteration progresses, many methods have been devised to choose nonoptimal, but, hopefully, 'good' approximations of the form (1), without necessitating the additional work and storage of GMRES. Examples were mentioned in section 1.2. We will not discuss the details of these methods here but note only that, unfortunately, there are no *a priori* results relating the convergence rate of these methods to that of the optimal GMRES. Some theorems may appear to provide such a relationship [8, 25, e.g.], but, inevitably, the established result either involves quantities generated during the computation that cannot be bounded *a priori*, or it allows for arbitrarily many look-ahead steps and so may not apply to a short recurrence. This lack of theoretical proof of near-optimality may be related to the previously mentioned result of Faber and Manteuffel — that, for most nonhermitian matrices A, there is no short recurrence for the optimal approximation from the space (1). This result deals only with a *single* recurrence and so does not rule out the possibility of finding the optimal approximation by using multiple short recurrences or other approaches that still require only $O(n)$ work and storage per iteration. The word 'optimal' here is taken to mean that the error is as small as possible in some inner product norm that *does not depend* on the initial vector x_0. One might be able to compute the approximation with the smallest error in some norm that *does* depend on the initial vector. In fact, Barth and Manteuffel have recently shown that *most* iterative methods generate approximations that are optimal in *some* norm [1]. Unfortunately, the norm is usually one that can be arbitrarily different from the 2-norm or the ∞-norm or any standard norm that is likely to be of interest! Perhaps the greatest remaining open problem in this field is to find a method that generates provably 'near-optimal' approximations in some standard norm, while still requiring only $O(n)$ work and storage (in addition to the matrix-vector multiplication) at each iteration — or to prove that such a method does not exist.

References

[1] T. Barth and T. Manteuffel, *Variable metric conjugate gradient methods*, in PCG '94 proceedings, Matrix Analysis and Parallel Computing, M. Natori and T. Nodera, eds., Keio University, Yokohama, 1994.

[2] V. Druskin, A. Greenbaum, and L. Knizhnerman, *Some uses of the Lanczos algorithm — and why it works!*, to appear.

[3] V. Druskin and L. Knizhnerman, *Error bounds in the simple Lanczos procedure for computing functions of symmetric matrices and eigenvalues*, Comput. Maths. Math. Phys., 31 (1991),

pp. 20–30.
[4] M. Engeli, T. Ginsburg, H. Rutishauser, and E. Stiefel, *Refined Iterative Methods for Computation of the Solution and the Eigenvalues of Self-adjoint Boundary Value Problems*, Birkhäuser-Verlag, Basel, Stuttgart, 1959.
[5] V. Faber, W. Joubert, M. Knill, and T. Manteuffel, *Minimal residual method stronger than polynomial preconditioning*, to appear in SIAM Jour. Matrix Anal. Appl.
[6] V. Faber and T. Manteuffel, *Necessary and sufficient conditions for the existence of a conjugate gradient method*, SIAM J. Numer. Anal., 21 (1984), pp. 352–362.
[7] R. Fletcher, *Conjugate gradient methods for indefinite systems*, in Proc. of the Dundee Biennial Conference on Numerical Analysis, G.A. Watson, ed., Springer-Verlag, New York, 1975.
[8] R. W. Freund and N. M. Nachtigal, *QMR: a quasi-minimal residual method for non-Hermitian linear systems*, Numer. Math. 60 (1991), pp. 315–339.
[9] A. Greenbaum, *Comparison of splittings used with the conjugate gradient algorithm*, Numer. Math., 33 (1979), pp. 181–194.
[10] A. Greenbaum, *Behavior of slightly perturbed Lanczos and conjugate gradient recurrences*, Lin. Alg. Appl. 113 (1989), pp. 7–63.
[11] A. Greenbaum and L. Gurvits, *Max-min properties of matrix factor norms*, SIAM J. Sci. Comput., 15 (1994), pp. 348–358.
[12] A. Greenbaum, V. Ptak, and Z. Strakoš, *Any nonincreasing convergence curve is possible for GMRES*, to appear in SIAM Jour. Matrix Anal. Appl.
[13] A. Greenbaum and Z. Strakoš, *Predicting the behavior of finite precision Lanczos and conjugate gradient computations*, SIAM J. Matrix Anal. Appl. 13, 1 (1992), pp. 121–137.
[14] A. Greenbaum and Z. Strakoš, *Matrices that generate the same Krylov residual spaces*, in Recent Advances in Iterative Methods, G. Golub, A. Greenbaum, and M. Luskin, eds., Springer-Verlag, New York, pp. 95–118, 1994.
[15] M. R. Hestenes and E. Stiefel, *Methods of conjugate gradients for solving linear systems*, J. Res. Nat. Bur. Standards, 49 (1952), pp. 409–435.
[16] W. A. Joubert, *A robust GMRES-based adaptive polynomial preconditioning algorithm for nonsymmetric linear systems*, SIAM J. Sci. Comput., 15 (1994), pp. 427–439.
[17] C. Lanczos, *An iteration method for the solution of the eigenvalue problem of linear differential and integral operators*, J. Res. Nat. Bur. Standards, 45 (1950), pp. 255–282.
[18] N. M. Nachtigal, L. Reichel, and L. N. Trefethen, *A hybrid GMRES algorithm for nonsymmetric linear systems*, SIAM J. Matrix Anal. Appl., 13 (1992), pp. 796-825.
[19] C. C. Paige, *Error Analysis of the Lanczos Algorithm for Tridiagonalizing a Symmetric Matrix*, J. Inst. Math. Appl. 18 (1976), pp. 341-349.
[20] C. C. Paige, *Krylov subspace processes, Krylov subspace mthods, and iteration polynomials*, in Proceedings of the Cornelius Lanczos International Centenary Conference, J. Brown, M. Chu, D. Ellison, and R. Plemmons, eds., SIAM, 1993.
[21] C. C. Paige and M. A. Saunders, *Solution of sparse indefinite systems of linear equations*, SIAM J. Numer. Anal. 11 (1974), pp. 197–209.
[22] Y. Saad and M. H. Schultz, *GMRES: A generalized minimal residual algorithm for solving nonsymmetric linear systems*, SIAM J. Sci. Stat. Comput., 7 (1986), pp. 856–869.
[23] P. Sonneveld, *CGS, a fast Lanczos-type solver for nonsymmetric linear systems*, SIAM J. Sci. Stat. Comput., 10 (1989), pp. 36–52.
[24] K. C. Toh, *GMRES vs. ideal GMRES*, to appear in SIAM Jour. Matrix Anal. Appl.
[25] C. H. Tong and Q. Ye, *A breakdown-free quasi-minimal residual algorithm for nonsymmetric linear systems*, to appear.
[26] L. N. Trefethen, *Approximation theory and numerical linear algebra*, in Algorithms for Approximation II, J. C. Mason and M. G. Cox, eds., Chapman and Hall, London, 1990.
[27] H. A. van der Vorst, *Bi-CGSTAB: A fast and smoothly converging variant of Bi-CG for the solution of nonsymmetric linear systems*, SIAM J. Sci. Comput., 13 (1992), pp. 631–644.

Chapter 8
Cholesky-based Methods
for Sparse Least Squares:
The Benefits of Regularization*

Michael A. Saunders[†]

Abstract

We study the use of black-box LDL^T factorizations for solving the augmented systems (KKT systems) associated with least-squares problems and barrier methods for linear programming (LP). With judicious regularization parameters, stability can be achieved for *arbitrary data* and *arbitrary permutations* of the KKT matrix.

This offers improved efficiency compared to implementations based on "pure normal equations" or "pure KKT systems". In particular, the LP matrix may be partitioned arbitrarily as $(A_s \ A_d)$. If $A_s A_s^T$ is unusually sparse, the associated "reduced KKT system" may have very sparse Cholesky factors. Similarly for least-squares problems if a large number of rows of the observation matrix have special structure.

Numerical behavior is illustrated on the villainous Netlib models *greenbea* and *pilots*.

1 Background

The connection between this work and Conjugate-Gradient methods lies in some properties of two CG algorithms, LSQR and CRAIG, for solving linear equations and least-squares problems of various forms. We consider the following problems:

(1)   *Linear equations*: $Ax = b$

(2)   *Minimum length*: $\min \|x\|^2$ subject to $Ax = b$

(3)   *Least squares*: $\min \|Ax - b\|^2$

(4)   *Regularized least squares*: $\min \|Ax - b\|^2 + \|\delta x\|^2$

(5)   *Regularized min length*: $\min \|x\|^2 + \|s\|^2$ subject to $Ax + \delta s = b$

where A is a general matrix (square or rectangular) and δ is a scalar ($\delta \geq 0$).

LSQR [17, 18] solves the first four problems, and incidentally the fifth, using essentially the same work and storage per iteration in all cases. The iterates x_k reduce $\|b - Ax_k\|$ monotonically.

CRAIG [4, 17] solves only compatible systems (1)–(2), with $\|x - x_k\|$ decreasing monotonically. Since CRAIG is slightly simpler and more economical than LSQR, it may sometimes be preferred for those problems.

---

*Partially supported by Department of Energy grant DE-FG03-92ER25117, National Science Foundation grant DMI-9204208, and Office of Naval Research grant N00014-90-J-1242.

[†]Systems Optimization Laboratory, Dept of Operations Research, Stanford University, Stanford, California 94305-4022 (mike@SOL-michael.stanford.edu).

To extend CRAIG to incompatible systems, we have studied problem (5): a compatible system in the combined variables (x, s). If $\delta > 0$, it is readily confirmed that problems (4) and (5) have the same solution x, and that both are solved by either the normal equations

(6) $$Nx = A^T b, \qquad N \equiv A^T A + \delta^2 I,$$

or the augmented system

(7) $$K \begin{pmatrix} s \\ x \end{pmatrix} = \begin{pmatrix} b \\ 0 \end{pmatrix}, \qquad K \equiv \begin{pmatrix} \delta I & A \\ A^T & -\delta I \end{pmatrix}.$$

The special form of CRAIG developed in [19] does not appear to have advantages over LSQR. However, a side-effect of that research has been to focus attention on system (7). Our aim in the remainder of the paper is to study *direct* methods for solving (6) and (7), with an emphasis on both stability and efficiency.

Some recent references exploring stability matters are Forsgren [7], Gill *et al.* [12], Vavasis [22] and Wright [23].

1.1 Notation

The following terms are used:

sqd	Symmetric quasi-definite, as in (8)
Cholesky factors	$PKP^T = LDL^T$, P a permutation, L lower triangular, D diagonal (possibly indefinite)
$\sigma_{\max}(A) = \|A\|$	Largest singular value of general matrix A
$\lambda_{\max}(K) = \|K\|$	Largest eigenvalue of symmetric matrix K
$\mathrm{Cond}(A) = \sigma_{\max}(A)/\sigma_{\min}(A)$	Condition number of general matrix A
$\mathrm{Cond}(K) = \lambda_{\max}(K)/\lambda_{\min}(K)$	Condition number of symmetric matrix K
$\mathrm{Econd}(K)$	Effective condition number of K when solving with unstable factors
ϵ	Floating-point precision (typically $\approx 10^{-16}$)

2 The Condition of N and K

The normal equations (6) are effective when N is sparse and reasonably conditioned. Since N is positive definite, it is well known that Cholesky factors $PNP^T = LL^T$ or LDL^T can be computed stably for all permutations P, and that P may therefore be chosen to preserve sparsity in L.

If N or its factors are *not* sparse or well-conditioned, the augmented system (7) may be of interest. In particular, it is better conditioned than the normal equations at times of importance—when A is ill-conditioned. By examining the eigenvalues of N and K, we obtain the following result for an arbitrary matrix A.

RESULT 1 ([19, §2]). *If $\delta > \sigma_{\min}(A)$, the condition numbers of N and K in (6)–(7) are as follows:* $\mathrm{Cond}(N) \approx (\|A\|/\delta)^2$, $\mathrm{Cond}(K) \approx \|A\|/\delta$.

If A comes from a sequence of increasingly ill-conditioned matrices, we see that regularization gives essentially *constant condition numbers*, and that K is much better conditioned than N. The implications for linear programming are pursued later.

A word of caution: If the degree of regularization is open to choice, δ should not be chosen "too small", since it could mean that $\|s\| \gg \|x\|$ in (7). Good accuracy in s may be accompanied by poor accuracy in x. With 16-digit precision, we recommend $\delta \geq 10^{-5}\|A\|$.

3 Cholesky on Quasi-definite Systems

In [21], Vanderbei introduced *symmetric quasi-definite* (sqd) matrices of the form

$$(8) \qquad K = \begin{pmatrix} H & A^T \\ A & -G \end{pmatrix}, \qquad H \text{ and } G \text{ symmetric positive definite.}$$

and advocated use of sparse Cholesky-type factors, $PKP^T = LDL^T$, for solving linear systems $Kz = d$. (See also [20]. Note that A is now transposed for the remainder of the paper.)

Since the Cholesky factors exist for all permutations, P may be chosen to maintain sparsity, as in the positive definite case. However, the usual excellent stability properties of Cholesky factors do not hold when K is indefinite. We must deal with this issue.

If a *stable* method were used to factorize K and solve $Kz = d$, the relative error in $\hat{z}$ (the computed z) would be bounded by an expression of the form

$$\|\hat{z} - z\|/\|z\| \le \epsilon \rho \operatorname{Cond}(K),$$

where ρ is a slowly-growing function of the dimension of K. If some other method is used to solve $Kz = d$, and if the relative error can be bounded by a similar expression with $\operatorname{Cond}(K)$ replaced by a quantity $\operatorname{Econd}(K)$, we define the latter to be an *effective condition number for K*.

An initial stability analysis of sqd systems follows from some results of Golub and Van Loan [13], as shown by Gill et al. [12].

RESULT 2 ([12, §4–5]). *If Cholesky factors are used to solve $Kz = d$, where K is the sqd matrix (8), an effective condition number is given by*

$$\omega(K) = \frac{\max\{\|A^T G^{-1} A\|, \|A H^{-1} A^T\|\}}{\|K\|}, \qquad \operatorname{Econd}(K) = \big(1 + \omega(K)\big) \operatorname{Cond}(K).$$

Typically, $\omega(K) \gg 1$ and we can omit the 1. For the regularized least-squares system (7), this gives the following result.

RESULT 3 ([19, §2.1]). *If Cholesky factors are used to solve $Kz = d$, where K is the matrix in (7), an effective condition number is $\operatorname{Econd}(K) \approx (\|A\|/\delta)^2$.*

Comparing with Result 1, we see that the effective condition of the augmented system K is the same as the true condition of the normal-equations matrix N (when N is ill-conditioned). Hence, if the Cholesky factors of K are sufficiently sparse, they may be preferable to those of N.

3.1 Iterative Refinement

Note that the right-hand side of $Kz = d$ in Result 3 is a general vector d. If iterative refinement is applied, errors in the computed corrections will again be governed by $\operatorname{Econd}(K)$ (and the accuracy of the right-hand side). Refinement is *not* effective with the associated normal equations unless the Cholesky factors are obtained from a QR factorization of A [2], which would usually be less efficient.

4 Regularized Linear Programs

Barrier methods for linear programming (e.g., [14]) give rise to sequences of increasingly ill-conditioned least-squares or sqd systems. Here we focus on the regularized LP problem discussed in [10, 11]:

(9)
$$\underset{x,s}{\text{minimize}} \quad c^T x + \tfrac{1}{2}\|\gamma x\|^2 + \tfrac{1}{2}\|p\|^2$$
$$\text{subject to} \quad Ax + \delta p = b, \quad l \le x \le u.$$

We assume that the problem has been scaled to satisfy $\|A\| \approx 1$. The scalars γ and δ are typically "small" ($\approx 10^{-4}$). Like all good regularization parameters, they ensure that an optimal primal and dual solution (x, π) exists and is bounded and unique for any values of the data (assuming $l \le u!$).

Each iteration of an "infeasible primal-dual barrier method" requires the solution of KKT systems of the form

(10) $\quad K \begin{pmatrix} \Delta x \\ -\Delta \pi \end{pmatrix} = \begin{pmatrix} w \\ r \end{pmatrix}, \quad K \equiv \begin{pmatrix} H & A^T \\ A & -\delta^2 I \end{pmatrix}, \quad H \equiv H_\mu + \gamma^2 I,$

where H_μ is diagonal and positive semi-definite. (Its elements change every iteration.) Zero diagonals of H_μ arise when there are free variables (with infinite bounds in l and u), but setting $\gamma > 0$ removes the common difficulty of forming the normal-equations matrix $N \equiv A H^{-1} A^T + \delta^2 I$.

4.1 Two Scalings of K

In the code PDQ1 [10, 11], we remove artificial ill-conditioning by scaling down the large diagonals of H, using a diagonal matrix D_1 with $(D_1)_{jj} = (\max\{H_{jj}, 1\})^{-1/2}$. This gives an equivalent system

(11) $\quad K_1 \begin{pmatrix} y \\ -\Delta \pi \end{pmatrix} = \begin{pmatrix} D_1 w \\ r \end{pmatrix}, \quad K_1 \equiv \begin{pmatrix} H_1 & D_1 A^T \\ A D_1 & -\delta^2 I \end{pmatrix},$

where $\Delta x = D_1 y$, $H_1 = D_1 H D_1$, $\|H_1\| = 1$, $\|H_1^{-1}\| \approx 1/\gamma^2$, $\|D_1\| = 1$, $\|AD_1\| \approx 1$, $\|K_1\| \approx 1$. The accuracy of the solution (and the need for iterative refinement) is based on the residuals for (11). Applying Result 2 to K_1 gives the following effective condition, as previously shown in [12].

RESULT 4 ([12, RESULT 6.1]). *If Cholesky factors are used to solve (11), the effective condition number is* $\text{Econd}(K_1) \approx \max\{1/\gamma^2, 1/\delta^2\} \text{Cond}(K_1)$.

On a typical LP problem, the barrier method generates 20 to 50 systems with $\text{Cond}(K_1)$ tending to increase as H changes. With $\gamma = \delta = 10^{-4}$, Result 4 seems to explain the viability of $P K_1 P^T = L D L^T$ factorizations within PDQ1, at least until the solution is approached (though it doesn't explain the success of Cholesky factorization in LOQO [20, 21], where P is chosen carefully without the help of regularization).

A difficulty with Result 4 is that $\text{Cond}(K_1)$ is not clearly bounded (though it may be moderate initially). A contribution of this paper is to convert (10) to a regularized least-squares system and obtain an effective condition number that is independent of H and therefore holds for all iterations of the barrier method. Scaling with $D_2 = H^{-1/2}$ and $1/\delta$ gives the equivalent system

$$(12) \qquad K_2 \begin{pmatrix} s \\ -\Delta\pi \end{pmatrix} = \begin{pmatrix} D_2 w \\ (1/\delta)r \end{pmatrix}, \qquad K_2 \equiv \begin{pmatrix} \delta I & D_2 A^T \\ A D_2 & -\delta I \end{pmatrix},$$

where $\Delta x = \delta D_2 s$, $\|D_2\| \approx 1/\gamma$, $\|AD_2\| \approx 1/\gamma$, $\|K_2\| \approx 1/\gamma$. Applying Result 1 gives $\text{Cond}(K_2) \approx \|D_2 A^T\|/\delta \approx 1/(\gamma\delta)$. Combining with Result 2 gives the following main result.

RESULT 5. *If Cholesky factors are used to solve (12), the effective condition number is* $\text{Econd}(K_2) \approx 1/(\gamma^2 \delta^2)$.

We see again that the effective condition of the augmented system K_2 is the same as the true condition of the normal-equations matrix $N = AD_2^2 A^T + \delta^2 I$ (when N is ill-conditioned). We may therefore favor K_2 if its factors are more sparse than those of N.

4.2 Scale Invariance

In practice, the numerical solution of $Kz = d$ using Cholesky factors with a given ordering is the *same* for all symmetric scalings of K (assuming overflow and underflow do not occur). Hence, Results 4 and 5 apply equally well to the original KKT system (10). The *best* of those results serves as a stability indicator.

Thus, we are free to choose *any ordering* P for the Cholesky factors of K in (10), as long as γ and δ are sufficiently large.

5 LBL^T Factorizations

Since the true condition of K_2 in (12) is only $1/(\gamma\delta) \approx 10^8$, implementations based on a *stable* factorization of K_2 should experience few numerical difficulties *regardless of the data*.

In PDQ1 we currently use the sparse indefinite solver MA47 [6], which performs somewhat better than its predecessor MA27 [5] in this context. These packages can form both LDL^T and LBL^T factorizations (with B block diagonal). The latter provide stability in the conventional sense, but tend to be less sparse. In general we request LDL^T as long as possible, and fall back on LBL^T factors with loose but increasingly strict tolerances if the KKT systems are not solved with sufficient precision (e.g., if γ and δ are too small). An alternative would be to continue with Cholesky factors after increasing γ and δ.

Fourer and Mehrotra [8] have implemented their own LBL^T factorizer and applied it to KKT systems closely related to K_2, using very loose stability tolerances. They would probably achieve similar success on K_2 itself, with LDL^T factors resulting if γ and δ are not too small.

6 Reduced KKT Systems

The KKT system (10) is often solved by forcing a block pivot on all of H and allowing a black-box Cholesky package to choose an ordering for the resulting normal equations. This is clearly stable if γ and δ are sufficiently large. However, several real-world models in [1, 15] illustrate the need for alternatives when AA^T or L are excessively dense.

Reduced KKT systems are formed by pivoting on *part* of H (say H_s). In PDQ1, an element of H_s is required to be "sufficiently large", and the associated column of A must be "sufficiently sparse". When the regularization parameters are large enough, the partition can be based solely on sparsity, as described next.

6.1 Dense Columns or Dense Factors

Let A be partitioned as $(A_s\ A_d)$, where the columns of A_d contain *ndense* or more nonzeros. Pivoting on the first part of H gives a *reduced KKT system* of the form

$$(13) \quad K_r \begin{pmatrix} \Delta\pi \\ \Delta x_d \end{pmatrix} = \begin{pmatrix} r \\ -w_d \end{pmatrix}, \quad K_r \equiv \begin{pmatrix} A_s H_s^{-1} A_s^T + \delta^2 I & A_d^T \\ A_d & -H_d \end{pmatrix}.$$

We then form a black-box factorization $PK_rP^T = LDL^T$. Acceptable values for *ndense* and P can be determined symbolically prior to the barrier iterations. For example, *ndense* $= 100$ might be successful in many cases (treating most columns of A as sparse), but if K_r or L exceed available storage, values such as $50, 20, 15, 10, 5, 1$ could be tried in turn. An intermediate value will probably be optimal.

6.2 Special Structures

In particular applications, A_s could be a large part of A with the property that $A_s A_s^T$ is unusually sparse. A benefit of regularization is that we are free to choose any such partition and then apply a black-box Cholesky package to the reduced matrix K_r. (Similarly for least-squares problems if many rows of the observation matrix have special structure.)

7 Numerical Results

To illustrate some effects, we report results from running the barrier code PDQ1 on two "eminent" LP problems from the Netlib collection [9]. The problems were scaled and then regularized ($\gamma, \delta > 0$). We requested 6 digits of accuracy in the regularized solution (x, π). (Iteration counts are about 10% greater when 8 digits are required.) Times are CPU seconds on a DEC Alpha 3000/400 workstation with about 16 digits of precision. MA47 was instructed to compute indefinite Cholesky factors of reduced KKT systems (13).

Table 1 shows the effect of varying γ and δ on problem *greenbea*, which can cause numerical difficulties in barrier methods without regularization (e.g., [21]).

1. With excessive regularization ($\gamma = \delta = 10^{-3}$), the final objective value is rather different from the optimum for the original problem. This is probably due to $\|x\|$ being unusually large.

2. With too little regularization ($\gamma = \delta = 10^{-6}$), Cholesky factorization of K_r becomes unstable. The last two iterations proceeded satisfactorily after MA47 switched to slightly more dense LBL^T factors (with stability tolerance 10^{-8}).

3. Most problems in the Netlib set have $\|x\| \approx 1$ after scaling, and give satisfactory solutions with $\gamma = \delta = 10^{-4}$. Implementations with a crossover to the simplex method should require relatively few simplex iterations to solve the original problem.

Table 2 shows the effect of varying the partition of A in forming reduced KKT systems (13) and their Cholesky factors. The values $\gamma = \delta = 10^{-4}$ gave reliable performance in all cases. $|K_r|$ is the number of terms summed to form K_r ($\times 1000$). $|L|$ is the number of nonzeros in L ($\times 1000$).

1. Problem *greenbea* is typical of "sparse" problems. Cholesky factors of N are significantly more sparse than for the full KKT system K.

2. Problem *pilots* contains a large number of *moderately* dense columns. The normal equations are reasonably efficient, but there is evidently scope for improvement with reduced KKT systems of various size (notably, *ndense*$= 5$).

TABLE 1
Barrier code PDQ1 on problem greenbea with various regularizations.

γ, δ	Itns	Final objective	$\|x\|$	$\|\pi\|$
10^{-3}	43	-6.948877×10^7	2700	25
10^{-4}	43	-7.246302×10^7	2800	54
10^{-5}	42	-7.246243×10^7	2800	1300
10^{-6}	44	-7.246264×10^7	6200	1100

TABLE 2
PDQ1 using Cholesky factors of various reduced KKT systems (13), including normal equations N and full system K. MA47 computes $PK_rP^T = LDL^T$. Normal equations are often efficient for sparse problems like greenbea. One of the reduced KKT systems (ndense = 5) is noticeably better for pilots.

	ndense	Cols in A_d	$|K_r|$	$|L|$	time	
greenbea	1000	0	102	113	54	N
	50	2	101	113	53	
	20	2	101	113	53	
	15	204	81	122	56	
	10	465	66	173	93	
	5	3833	37	272	131	
	1	5495	39	295	140	K
pilots	1000	0	530	230	187	N
	50	77	352	235	169	
	20	679	113	280	174	
	15	975	79	276	162	
	10	1432	53	290	160	
	5	2121	44	300	148	
	1	4657	48	371	170	K

8 Least Squares with Bounds

As an example of augmented systems that are regularized "naturally", we consider least-squares problems with bounded variables. Barrier methods generate systems that are increasingly ill-conditioned (as in the LP case), but again the systems can be solved with LDL^T factors, as we now show.

For simplicity, let the problem be

$$\min \|Ax - b\|^2 \quad \text{subject to} \quad x \geq 0 \tag{14}$$

and consider the function $F(x, \mu) = \frac{1}{2}\|Ax-b\|^2 - \mu \sum \ln x_j$, where μ is the barrier parameter ($\mu > 0$). A primal barrier method applies Newton's method to minimize $F(x,\mu)$ as a function of x. Each iteration requires the solution of

$$(A^TA + \mu X^{-2})\Delta x = A^T r + \mu X^{-1} e, \tag{15}$$

where x is the current estimate ($x > 0$), $X = \text{diag}(x_j)$, $r = b - Ax$, and e is a vector of 1's. In some applications, it may be best to treat this system directly. Otherwise, we may write it as the least-squares problem

$$\min_t \left\| \begin{pmatrix} AX \\ \delta I \end{pmatrix} t - \begin{pmatrix} r \\ \delta e \end{pmatrix} \right\|^2, \qquad (16)$$

where $\delta = \sqrt{\mu}$ and $\Delta x = Xt$. The solution is given by the augmented system

$$K \begin{pmatrix} s \\ t \end{pmatrix} = \begin{pmatrix} r \\ -\delta e \end{pmatrix}, \qquad K \equiv \begin{pmatrix} \delta I & AX \\ XA^T & -\delta I \end{pmatrix}, \qquad (17)$$

whose regularization parameter is *prescribed* in terms of the barrier parameter μ. Although μ tends to zero as the barrier method converges, it should comfortably satisfy $\mu \geq \|b\|^2 \epsilon$. From Result 3, we have

$$\mathrm{Econd}(K) \approx (\|AX\|/\delta)^2 \approx (\|Ax\|/\delta)^2 \leq \|b\|^2/\mu \leq 1/\epsilon.$$

Hence, LDL^T factors of K should be sufficiently stable throughout.

9 Conclusions

Although sparse LBL^T packages are available for indefinite systems, they are inevitably more complex than Cholesky codes. Regularization expands the latter's applicability. Some final comments follow:

1. Sparse Cholesky codes are often implemented to compute LL^T factors on the assumption that they will be applied to positive definite systems. With little change they could produce LDL^T factors and allow D to have both positive and negative elements. MA27 and MA47 already do so.

2. We advocate the use of such black-box packages (and any new ones that come along) for solving sparse least-squares problems with regularization. Barrier methods for linear programming are a natural application. The parallelized Cholesky solver used in [16] is a promising candidate for wider use.

3. The same LDL^T packages may be applied to normal equations N, to full KKT systems K (10), or to the spectrum of reduced KKT matrices K_r (13). Regularization ensures adequate stability in all cases, allowing the choice to be based solely on the sparsity of K_r and its factors.

4. Reduced KKT systems promise efficiency in special cases where $A = (A_s \; A_d)$, if A_s is a large part of A and $A_s A_s^T$ is unusually sparse.

5. Similar techniques apply to bound-constrained least-squares problems. If a linear program is suspected of being infeasible, the approach of Section 8 might provide a useful "best" solution. (Alternatively, the primal-dual barrier method of Section 4 could be applied with $\delta = 1$. This has proved more effective in recent experiments on large, dense, infeasible LP problems [3].)

6. With today's 64-bit machines, the range of permissible regularization parameters is somewhat narrow. If higher precision becomes commonplace, the solution of quasi-definite systems (via Cholesky factorization) will be an important beneficiary.

7. So too will the solution of *unsymmetric* systems $Ax = b$ (via Cholesky factorization of system (7), with iterative refinement to minimize the effect of δ).

References

[1] R. E. Bixby, *Progress in linear programming*, ORSA J. on Computing, 6(1) (1994), pp. 15–22.
[2] Å. Björck, *Stability analysis of the method of seminormal equations for linear least squares problems*, Linear Alg. and its Appl., 88/89 (1987), pp. 31–48.
[3] S. Chen (private communication), Computational experiments on infeasible LP problems arising from Basis Pursuit methods for de-noising images, Dept of Statistics, Stanford University, Stanford, CA, 1995.
[4] J. E. Craig, *The N-step iteration procedures*, J. Math. and Phys., 34(1) (1955), pp. 64–73.
[5] I. S. Duff and J. K. Reid, *The multifrontal solution of indefinite sparse symmetric linear systems*, ACM Trans. Math. Softw., 9 (1983), pp. 302–325.
[6] I. S. Duff and J. K. Reid, *Exploiting zeros on the diagonal in the direct solution of indefinite sparse symmetric linear systems*, ACM Trans. Math. Softw., to appear.
[7] A. Forsgren, *On linear least-squares problems with diagonally dominant weight matrices*, Report TRITA-MAT-1995-OS2, Dept of Mathematics, Royal Institute of Technology, Stockholm, Sweden, 1995.
[8] R. Fourer and S. Mehrotra, *Solving symmetric indefinite systems in an interior-point method for linear programming*, Math. Prog., 62 (1993), pp. 15–39.
[9] D. M. Gay, *Electronic mail distribution of linear programming test problems*, Mathematical Programming Society COAL Newsletter, 13 (1985), pp. 10–12.
[10] P. E. Gill, W. Murray, D. B. Ponceleón, and M. A. Saunders, *Solving reduced KKT systems in barrier methods for linear and quadratic programming*, Report SOL 91-7, Dept of Operations Research, Stanford University, CA, 1991.
[11] ———, *Solving reduced KKT systems in barrier methods for linear programming*, In G. A. Watson and D. Griffiths (eds.), *Numerical Analysis 1993*, Pitman Research Notes in Mathematics 303, Longmans Press, pp. 89–104, 1994.
[12] P. E. Gill, M. A. Saunders and J. R. Shinnerl, *On the stability of Cholesky factorization for quasi-definite systems*, SIAM J. Mat. Anal., to appear.
[13] G. H. Golub and C. F. Van Loan, *Unsymmetric positive definite linear systems*, Linear Alg. and its Appl., 28 (1979), pp. 85–98.
[14] I. J. Lustig, R. E. Marsten and D. F. Shanno, *On implementing Mehrotra's predictor-corrector interior point method for linear programming*, SIAM J. Optim., 2 (1992), pp. 435–449.
[15] ———, *Interior point methods for linear programming: Computational state of the art*, ORSA J. on Computing, 6(1) (1994), pp. 1–14.
[16] I. J. Lustig and E. Rothberg, *Gigaflops in linear programming*, Technical Report, 1995.
[17] C. C. Paige and M. A. Saunders, *LSQR: An algorithm for sparse linear equations and sparse least squares*, ACM Trans. Math. Softw., 8(1) (1982), pp. 43–71.
[18] ———, *Algorithm 583. LSQR: Sparse linear equations and least squares problems*, ACM Trans. Math. Softw., 8(2) (1982), pp. 195–209.
[19] M. A. Saunders, *Solution of sparse rectangular systems using LSQR and CRAIG*, Report SOL 94-4, Dept of Operations Research, Stanford University, CA, 1994. To appear in BIT.
[20] R. J. Vanderbei, *LOQO: An interior point code for quadratic programming*, Report SOR-94-15, Dept of Statistics and Operations Research, Princeton University, Princeton, NJ, 1994.
[21] ———, *Symmetric quasi-definite matrices*, SIAM J. Optim., 5(1) (1995), pp. 100–113.
[22] S. A. Vavasis, *Stable numerical algorithms for equilibrium systems*, SIAM J. Mat. Anal., 15 (1994), pp. 1108–1131.
[23] S. Wright, *Stability of linear algebra computations in interior-point methods for linear programming*, Report MCS-P446-0694, Mathematics and Computer Science Division, Argonne National Laboratory, Argonne, IL, 1994.

Chapter 9
Memory Retention in Conjugate Gradient Methods

Laurence Dixon*

Abstract

This paper was inspired by considering the problem of minimising a quadratic function subject to bounds on the variables. A method was required that retained the property of the conjugate gradient method for the unbounded problem, namely finite termination without using a matrix store. Investigation of this problem led to a reconsideration of the properties of the conjugate gradient method. Two sets of new results are presented. The first is concerned with the properties needed if the iteration is commenced at a point where prior information is known and is to be retained. The second considers the behaviour after a steepest descent step. Such a step is sometimes advocated to prove convergence in the presence of rounding error but destroys the finite termination property. A way of introducing such a step while retaining finite termination is presented.

1 Introduction

In this paper we will be concerned with the behaviour of the conjugate gradient method in non-standard circumstances. The conjugate gradient method is frequently used to solve a set of symmetric positive definite linear equations

(1) $$Ax = b.$$

This is well known to be equivalent to solving the problem of minimising the quadratic function Q given by

(2) $$Q = \frac{1}{2}x^T Ax - b^T x.$$

The challenge that inspired this paper is that of minimising Q subject to non-negativity bounds on the variables

(3) $$\min\ Q(x) \quad \text{s.t.} \quad x_i \geq 0 \quad i = 1,\ldots,n$$

by a conjugate gradient style of method. The conjugate gradient method is well known to solve (2) in a finite number of steps with the advantage of not requiring a matrix store for A. When the product Ap can be calculated efficiently without it, the method then only requires a small number of vector stores. The challenge is still not solved, but the investigation has led to a number of new results which are presented in this article.

---

*Emeritus Professor and Senior Research Fellow, Numerical Optimisation Centre, School of Information Sciences, University of Hertfordshire, Hatfield, UK.

2 The Standard Conjugate Gradient Method

The conjugate gradient method for solving (1) and (2) depends on the construction of a set of directions that are required by the theory but not stored. There are many variations of the conjugate gradient approach: the first originated with Hestenes and Stiefel [3]. It is iterative and starts at an arbitrary point x_1. At x_k the gradient of Q is denoted by g_k so

$$g_k = Ax_k - b. \tag{4}$$

At x_k the method generates the next conjugate direction by

$$p_k = -g_k + \beta p_{k-1}, \quad p_0 = 0 \tag{5}$$

where β is chosen so $p_{k-1}^T A p_k = 0$, so

$$\beta = \frac{p_{k-1}^T A g_k}{p_{k-1}^T A p_{k-1}}. \tag{6}$$

This formula for β can be simplified to give alternative expressions used in the Fletcher-Reeves [2] and Polak-Ribière [9] methods. In exact arithmetic these have the same properties and generate the same points as (6) on quadratic functions. The differences become more significant when the method is extended to non-quadratic functions.

When the new direction has been calculated by (5) and (6) a line search is undertaken so

$$x_{k+1} = x_k + \alpha p_k \tag{7}$$

$$g_{k+1} = g_k + \alpha A p_k \tag{8}$$

where α is chosen so $p_k^T g_{k+1} = 0$, i.e.

$$\alpha = \frac{p_k^T g_k}{p_k^T A p_k}. \tag{9}$$

The formulae for α and β can not involve division by zero if p_k is non-zero and A is positive definite. The vector $y_k = A p_k$ is of course only calculated once for each value of k.

The conjugate gradient method has a number of advantages especially when the matrix A is large and sparse as A is only involved in the calculation of y_k and no other matrix store is used. The vectors x, p, g, y can all be overwritten in the next iteration so only four vector stores are required. In exact arithmetic the method possesses finite termination to the solution where $g = 0$ in a number of iterations not greater than the number of distinct eigenvalues of A. In floating point arithmetic the iteration is usually terminated when $||g_k|| \leq \epsilon$ for some suitably small value of ϵ.

Unfortunately the method is numerically unstable in the presence of rounding errors and due to the magnification of rounding errors on problems with ill-conditioned A these theoretical properties are lost. This is discussed in Section 4.

The usual proofs that the conjugate gradient method generates directions p_k and g_k with the properties that

$$p_k^T A p_j = 0 \quad \text{for all } j < k \tag{10}$$

$$g_k^T g_j = 0 \quad \text{and} \quad g_k^T p_j = 0 \quad \text{for all } j < k \tag{11}$$

are inductive and start at $k = 1$. In this paper we will consider each iteration separately.

3 Non-inductive Results

THEOREM 3.1. *If at x_k the directions p_k and g_k satisfy four conditions for a subspace S_k, namely, if $s \in S_k$, then*

$$C1:\ g_k^T s = 0, \quad C2:\ p_k^T As = 0, \quad C3:\ p_k \in S_k + g_k, \quad C4:\ As \in S_k + g_k$$

and if $x_{k+1}, g_{k+1}, p_{k+1}$ and S_{k+1} are defined by

$$D1:\ x_{k+1} = x_k + \alpha p_k, \quad D2:\ g_{k+1} = g_k + \alpha A p_k \text{ with } \alpha \text{ s.t. } p_k^T g_{k+1} = 0,$$

$$D3:\ p_{k+1} = -g_{k+1} + \beta p_k \text{ with } \beta \text{ s.t. } p_k^T A p_{k+1} = 0, \text{ and} \quad D4:\ S_{k+1} = S_k + g_k$$

then g_{k+1} and p_{k+1} satisfy the following four properties for all $s \in S_{k+1}$

$$P1:\ g_{k+1}^T s = 0, \quad P2:\ p_{k+1}^T As = 0, \quad P3:\ p_{k+1} \in S_{k+1} + g_{k+1}, \quad P4:\ As \in S_{k+1} + g_{k+1}$$

It will be noted that the four properties P1–P4 satisfied by the new directions are identical to the four conditions C1–C4 satisfied by the original directions and that the conditions for induction follow. It will also be noted that the definitions of p_{k+1} and g_{k+1} are the same as those used in the conjugate gradient method but that the method of obtaining S_k is no longer specified. The result is therefore more general than the standard inductive proof as we can now start at any point with any subspace S and direction p that satisfy conditions C1–C4 with respect to g.

Proof. Property P1 follows as if $s \in S_{k+1}$ then it consists of a component along p_k and a component in S_k, $p_k^T g_{k+1} = 0$ from the definition of α in D2, the other component is zero by C1 and C2.

Property P2 follows as now the component along p_k contributes zero by the definition of β in D3, while if $s \in S_k$, as $s^T A p_{k+1} = -s^T A g_{k+1} + \beta s^T A p_k$, the first term is zero by C4 and P1 and the second term is zero by C2.

Property P3 follows directly from C3, D3 and D4.

Property P4 follows as the component in S_k gives a component of As in S_{k+1} by C4 and D4, while from D2 the component $Ap_k = (g_{k+1} - g_k)/\alpha$ and $g_k \in S_{k+1}$ from D4.

Similar theorems can be stated for other conjugate gradient methods. Nazareth [7] introduced the three-term formula, with additional properties given in Nocedal [8]:

$$(12) \qquad p_{k+1} = Ap_k + \beta p_k + \gamma p_{k-1}$$

where β and γ are chosen so the $p_k^T A p_{k+1} = 0$, and $p_{k-1}^T A p_{k+1} = 0$.

THEOREM 3.2. *If the directions p_{k+1} and p_k and the directions in subspace S_{k+1} satisfy the following four conditions for all $s \in S_{k-1}$*

$$C1:\ s^T Ap_k = 0, \quad C2:\ s^T Ap_{k-1} = 0, \quad C3:\ p_k^T Ap_{k-1} = 0, \text{ and} \quad C4:\ As \in S_{k-1} + p_{k-1}$$

and if we define

$$D1:\ p_{k+1} \text{ by (12) and } D2:\ S_k = S_{k-1} + p_{k-1}$$

then for all $s \in S_k$, the following four properties hold

$$P1:\ s^T Ap_{k+1} = 0, \quad P2:\ s^T Ap_k = 0, \quad P3:\ p_k^T Ap_{k+1} = 0 \text{ and} \quad P4:\ As \in S_k + p_k$$

If in addition $g_k^T s = 0$ for all $s \in S_k$ and g_{k+1} is obtained by an exact line search along p_k, then $g_{k+1}^T s = 0$ for all $s \in S_k + p_k$.

The proof is very similar to that of Theorem 3.1 and is therefore omitted. The identical nature of the properties P1–P4 to the conditions C1–C4 is obvious as is the decoupling between the conjugate and gradient sets. The possibility of switching between the two formulae in one iteration and even of comparing nominally identical x_k calculated by both to obtain better values and to indicate when rounding errors become significant now becomes possible.

4 The Effect of Steepest Descent Steps

The investigation of the effect of a steepest descent step was prompted by the need to leave active bounds when solving problem (3). The insertion of a regular series of steepest descent steps into the non-quadratic versions of the conjugate gradient algorithm has a long history. If such steps are inserted regularly, then it can be shown that the magnitude of the gradient is reduced below ϵ in a finite number of steps, as the bound in the reduction in the function value in those steps is independent of the prior rounding error. It is always hoped and expected that the intervening conjugate gradient steps will significantly reduce the number of steps required to achieve this. As the conjugate gradient steps are unstable with respect to rounding error growth it can also be argued that there is a maximum number of such steps that can sensibly be used.

The analysis in [1] of the effect of a steepest descent step at x_j on the subset of p_i, g_i $i < j$ that retain the conjugacy, orthogonality properties with respect to p_k, g_k $k > j$, indicated that one value of i was lost at each subsequent iteration so that by iteration $2j$ all memory of the iterations with $i < j$ is lost. This seems a most unfortunate situation.

Two methods are proposed in [1] to improve it. Let P_k be the set of directions $p_1, \ldots, p_k$ and G_k be the set of gradients $g_1, \ldots, g_k$. Now if a steepest descent step is inserted into the standard conjugate gradient iteration at step j and each subsequent p_k is defined by

$$(13) \qquad p_k = -g_k + \beta p_{k-1} + \gamma p_{j-2}$$

where β is chosen to make $p_k^T A p_{k-1} = 0$ and γ to make $p_k^T A p_{j-2} = 0$ then the following theorem holds.

THEOREM 4.1. *If after a steepest descent step at iteration j subsequent directions are defined by (13) then if $S = P_k - p_{j-1}$ and $s \in S$*

$$P1: \ s^T g_k = 0 \quad and \quad P2: \ s^T A p_k = 0.$$

Properties P1 and P2 also apply if $S = G_k - g_{j-1}$.

This is an ideal result, unfortunately it does not extend to a second steepest descent step. If a second steepest descent step is introduced at iteration l then, if the orthogonality properties $g_{l+1}^T g_i = 0$ are to be retained, subsequent iterations must involve a two dimensional search [1].

5 Two Dimensional Searches

If we combine equations (5) and (7) we obtain

$$(14) \qquad x_{k+1} = x_k - \alpha g_k + \alpha \beta p_{k-1}$$

where, from (11), α and β are chosen so $g_{k+1}^T g_k = 0$ and $g_{k+1}^T p_{k-1} = 0$. These are the conditions for x_{k+1} to be the minimum in the two dimensional space containing x_k and spanned by g_k and p_{k-1}. The concept of a two-dimensional search in conjugate gradient theory and practise was introduced by Liu and Storey [6] and Hu and Storey [4] for very different reasons. They were mainly concerned with inaccurate line searches. Related results are given in [5] and [10]. The solution of the two dimensional search can be performed in many different ways and each author has done it differently. The one described below is yet another variant. The two dimensional search can be carried out by two lines searches, the first along the steepest descent direction $-g_k$ and the second along the direction conjugate to g_k in the two dimensional space. The first guarantees the reduction of $\|g_k\|$ below ϵ in a finite number of steps even in the presence of rounding error, and the second always reduces Q below the steepest descent value even with rounding error. The other two-dimensional searches would have this property in exact arithmetic but not necessarily after rounding error. In exact arithmetic the two-dimensional search generates the same points as the standard method. The defining equations are

(15) $$x_s = x_k - \alpha_s g_k \text{ with } \alpha_s \text{ chosen so } g_s^T g_k = 0$$

(16) $$p_s = -g_k + \beta_s p_{k-1} \text{ with } \beta_s \text{ chosen so } p_s^T A g_k = 0$$

(17) $$x_{k+1} = x_s + \alpha p_s \text{ with } \alpha \text{ chosen so } g_{k+1}^T p_s = 0$$

(18) $$p_k = x_{k+1} - x_k.$$

A non-inductive theorem for the two-dimensional search variant can also be stated.
THEOREM 5.1. *If at x_k the directions g_k and p_{k-1} and the subspace S_{k-1} satisfy the following conditions for all $s \in S_{k-1}$*

$$C1: s^T g_k = 0, \quad C2: s^T A p_{k-1} = 0, \quad C3: s^T A g_k = 0, \quad C4: g_k^T p_{k-1} = 0,$$

$$C5: As \in S_{k-1} + g_k \text{ and } C6: Ap_{k-1} \in S_{k-1} + g_k$$

and if we define

$$D1: x_{k+1}, g_{k+1} \text{ by (17)}, \quad D2: p_k \text{ by (18) and } D3: S_k = S_{k-1} + p_{k-1}$$

then for all $s \in S_k$ we have

$$P1: s^T g_{k+1} = 0, \quad P2: s^T A p_k = 0, \quad P3: s^T A g_{k+1} = 0, \quad P4: p_k^T g_{k+1} = 0,$$

$$P5: As \in S_k + g_{k+1} \text{ and } P6: Ap_k \in S_k + g_{k+1}$$

The proof of Theorem 5.1 is again similar to Theorem 3.1 and is therefore omitted. As presented Theorem 3.1 imposes conditions on p_k, the direction after x_k, while Theorem 5.1 imposes conditions on p_{k-1}, the direction before x_k. A further theorem for the standard method imposing conditions on p_{k-1} rather than p_k could also be stated.

6 Conclusions

In this paper a series of theorems have been stated giving the conditions that must apply at x_k between the gradient g_k, a direction p and a subspace S if three variants of the conjugate gradient method are to recreate these conditions at x_{k+1}. The behaviour of the conjugate gradient method after a steepest descent step is discussed and a new variant of the two dimensional search method introduced.

References

[1] L. C. W. Dixon, *On the restarted conjugate gradient method*, Tech. Report 295, NOC, University of Hertfordshire, UK, 1995.
[2] R. Fletcher and C. M. Reeves, *Function minimization by conjugate gradients*, Computer Journal, 7 (1964), pp. 149–154.
[3] M. R. Hestenes and E. Steifel, *Method of conjugate gradients for solving linear systems*, J. Res. Nat. Bus. Standards, 48 (1952), pp. 409–436.
[4] Y. F. Hu and C. Storey, *Efficient generalised conjugate gradient algorithms: 2 implementation*, JOTA, 69 (1991), pp. 139–152.
[5] J. Koko, *Generalised conjugate gradient algorithms with quasi-Newton approximations*, Tech. Report, Universite Blaise Pascal, Paris, France, 1994.
[6] Y. Liu and C. Storey, *Efficient generalised conjugate gradient algorithms: 1 theory*, JOTA, 69 (1991), pp. 129–137.
[7] J. L. Nazareth, *A conjugate gradient algorithm without line searches*, JOTA, 23 (1977), pp. 373–387.
[8] J. Nocedal, *On the Method of Conjugate Gradients for Function Minimisation*, PhD dissertation, Rice University, 1978.
[9] E. Polak and G. Ribière, *Note sur la convergence de methodes de directions conjuguees*, Rev. Francaise d'Informatique Recherche Oper., 16 (1969), pp. 35–43.
[10] L. E. Stephenson and J. L. Nazareth, *Successive affine reduction and a modified Newton method*, Tech. Report 95-1, Washington State University, Pullman, USA, 1995.

Chapter 10
Conjugate Gradient Algorithms Using Multiple Recursions*

Teri L. Barth[†]   Thomas A. Manteuffel[‡]

Abstract

Much is already known about when a conjugate gradient method can be implemented with short recursions for the direction vectors. The work done in 1984 by Faber and Manteuffel gave necessary and sufficient conditions on the iteration matrix A, in order for a conjugate gradient method to be implemented with a single recursion of a certain form. However, this form does not take into account all possible recursions. This became evident when Jagels and Reichel ([5],[6]) demonstrated that the class of matrices for which a practical conjugate gradient algorithm exists, can be extended to include unitary and shifted unitary matrices. The implementation uses short double recursions for the direction vectors. This motivates the study of multiple recursion algorithms. In this paper, we show that the conjugate gradient method for unitary and shifted unitary matrices can be implemented using a single short term recursion of a special type called an (ℓ, m) recursion with $\ell, m \leq 1$. We then examine the class of matrices for which a conjugate gradient method can be carried out using a general (ℓ, m) recursion.

1 Introduction
1.1 Motivation

A conjugate gradient (CG) method for the solution of an $N \times N$ linear system of equations

$$A\underline{x} = \underline{b},$$

is defined with respect to a Hermitian positive definite (HPD) inner product matrix B, and the matrix A. We denote a conjugate gradient method by $CG(B, A)$. These methods produce iterates that are uniquely defined as follows:

$$\begin{aligned}\underline{x}_{i+1} &= \underline{x}_i + \underline{z}_i, & \underline{z}_i &\in \mathcal{K}_{i+1}(\underline{r}_0, A), \\ \underline{e}_{i+1} &\perp_B \underline{z}, & \forall \underline{z} &\in \mathcal{K}_{i+1}(\underline{r}_0, A),\end{aligned}$$

where

$$\mathcal{K}_{i+1}(\underline{r}_0, A) = \mathrm{sp}\{\underline{r}_0, A\underline{r}_0, ..., A^i\underline{r}_0\},$$

is the Krylov subspace of dimension $i+1$ generated by the initial residual $\underline{r}_0$, and the matrix A, the error at step $i + 1$ is denoted by $\underline{e}_{i+1} = \underline{x} - \underline{x}_{i+1}$, and $\perp_B$ represents orthogonality

---

*This work was sponsored by the National Science Foundation under grant number DMS-8704169, and the Department of Energy under grant number DE-FG03-93ER25217.

[†]Department of Mathematics, University of Colorado at Denver, Denver, CO.

[‡]Program in Applied Mathematics, Campus Box 526, University of Colorado at Boulder, Boulder, CO 80309-0526.

in the B-inner product (c.f.[1]). $CG(B, A)$ is implemented via the construction of a B-orthogonal basis $\{\underline{p}_j\}_{j=0}^{i}$ for $\mathcal{K}_{i+1}(\underline{r}_0, A)$. The $\underline{p}_j$'s are called direction vectors. There are many algorithms for implementing $CG(B, A)$, for example: Orthodir, Orthomin, Orthores, and GMRES. A result from 1984 by Faber and Manteuffel states that the class of matrices for which a conjugate gradient method can be implemented via a single short s-term recursion of the form

$$\underline{p}_{i+1} = A\underline{p}_i - \sum_{j=i-s+2}^{i} \sigma_{i,j}\underline{p}_j, \tag{1}$$

for the direction vectors, is limited to matrices that are B-normal(s-2) (see Section 4.2), or to matrices with only a small number of distinct eigenvalues [3]. For these matrices, Orthodir and GMRES are two implementations of $CG(B, A)$ that have recursions of the form given by (1). Joubert and Young [7] extended this result, showing that the Orthomin and Orthores implementations of $CG(B, A)$ also yield short recursions for the direction vectors for B-normal(s-2) matrices.

The work done by Gragg [4] demonstrates that for a unitary matrix U, an orthonormal basis for $\mathcal{K}_{i+1}(\underline{r}_0, U)$ can be constructed using a pair or recurrence formulas. Shifted unitary matrices have the form

$$A = \rho U + \xi I, \tag{2}$$

where U is unitary, I is the identity matrix, and ρ and ξ are complex scalars. Jagels and Reichel ([5],[6]) observed that for matrices of this form, $\mathcal{K}_{i+1}(\underline{r}_0, A) = \mathcal{K}_{i+1}(\underline{r}_0, U)$, and from this they show how to use the Gragg algorithm (see Section 2.1) to construct an efficient minimal residual algorithm. This is an example of a class of matrices that is not B-normal(s-2), for small s, but for which a conjugate gradient method can be implemented using a pair of short recursions. Thus, by considering double recursions, the class of matrices for which a practical conjugate gradient algorithm exists can be extended to include unitary and shifted unitary matrices. In this paper, we will explore multiple recursions further.

Section 2 reviews the related work done by Gragg, Jagels and Reichel. In Section 3, we give an alternative minimal residual algorithm for shifted unitary matrices. Next, we will show that the pair of recurrence formulas used to construct an orthonormal basis for $\mathcal{K}_{i+1}(\underline{r}_0, A)$, when A is of the form (2), can be rewritten as a single recursion of the form

$$\underline{p}_{i+1} = \sum_{j=i-m}^{i} \beta_{i,j} A\underline{p}_j - \sum_{j=i-\ell}^{i} \sigma_{i,j}\underline{p}_j, \tag{3}$$

where $\ell = m = 1$. Section 4 considers general recursions of this form, where ℓ, m can be any integers ≥ 0. If a conjugate gradient method can be implemented with a short recursion of the form (3), we say $A \in \mathcal{CG}_B(\ell, m)$. Sufficient conditions on the matrix A are given for $A \in \mathcal{CG}_B(\ell, m)$. In [2] it is shown that for $\ell, m \leq 2$ these are also necessary conditions. Necessary conditions for $\ell, m > 2$ remain an open question. Section 5 summarizes these results. We conclude with a few remarks about work still in progress.

1.2 Notation

We will make use of the following notation throughout this paper.

$\mathcal{R}^n, \mathcal{C}^n$	—	Vector spaces of real and complex $n-$tuples.
$\mathcal{R}^{n \times m}, \mathcal{C}^{n \times m}$	—	Vector spaces of real and complex $n \times m$ matrices.
$\mathcal{P}_k$	—	Space of polynomials of degree at most k.
$\mathcal{P}_k^o$	—	Space of polynomials, $p_k(\lambda) \in \mathcal{P}_k$, such that $p_k(0) = 1$.
$\mathcal{K}, \mathcal{V}, \ldots$	—	Other calligraphic letters denote subspaces of $\mathcal{R}^n$ or $\mathcal{C}^n$.
$A, B, \ldots$	—	Upper case Roman and Greek letters denote matrices.
$\underline{x}, \underline{p}, \ldots$	—	Underlined lower case Roman and Greek letters denote vectors.
$\alpha, \beta, \ldots$	—	Lower case Roman and Greek letters denote scalars.
$\langle \cdot, \cdot \rangle, \|\cdot\|$	—	Euclidean inner product on $\mathcal{C}^n$ and induced norm.
$\langle B\cdot, \cdot \rangle, \|\cdot\|_B$	—	$B-$inner product on $\mathcal{C}^n$ and induced $B-$norm.
A^*	—	Euclidean adjoint of A, $A^* = \bar{A}^T$.
$A^\dagger$	—	$B-$adjoint of A.
$\text{sp}\{\underline{x}_j\}$	—	The linear span of the vectors $\underline{x}_j$.
$\Sigma(A)$	—	Spectrum of A.
$\mathcal{K}_\ell(\underline{r}_0, A)$	—	$\text{sp}\{\underline{r}_0, A\underline{r}_0, ..., A^{\ell-1}\underline{r}_0\}$ Krylov subspace of dimension ℓ.
$d(\underline{r}_0, A)$	—	Maximum dimension of Krylov subspace generated by $\underline{r}_0$ and A.
$d(A)$	—	Degree of the minimal polynomial of A.

2 Unitary and shifted unitary matrices

2.1 The Gragg algorithm for unitary matrices

It was shown by Gragg [4] that if U is an $N \times N$ unitary matrix, an orthonormal basis for $\mathcal{K}_\ell(\underline{v}_0, U)$ can be constructed with two short recursions. Suppose we let $\underline{v}_0 = \frac{\underline{r}_0}{\|\underline{r}_0\|}$, the normalized initial residual, and use a Gram-Schmidt process to construct an orthonormal basis $\{\underline{v}_j\}_{j=0}^{\ell-1}$ for $\mathcal{K}_\ell(\underline{r}_0, U)$. Denote the $N \times \ell$ matrix

$$V_\ell = \begin{bmatrix} | & | & & | \\ \underline{v}_0 & \underline{v}_1 & \cdots & \underline{v}_{\ell-1} \\ | & | & & | \end{bmatrix},$$

and suppose that $d(\underline{r}_0, U) = N$, where $d(\underline{r}_0, U)$ is the dimension of the maximum Krylov subspace generated by $\underline{r}_0$ and U. At step N the recursions for the $\underline{v}_j$'s may be written in matrix notation as

(4) $$UV_N = V_N H_N,$$

where H_N is an $N \times N$ upper Hessenberg matrix. Since U is unitary, and by construction V_N is unitary, it follows from (4) that $V_N H_N$ is unitary and

$$(V_N H_N)^* V_N H_N = H_N^* H_N = I_N,$$

the $N \times N$ identity matrix. Therefore, H_N is a unitary, upper Hessenberg matrix. A QR factorization of H_N yields

$$H_N = Q_N R_N,$$

where Q_N is $N \times N$ unitary, and R_N is $N \times N$ upper triangular. Since $R_N = Q_N^* H_N$, it follows that R_N is also unitary. R_N being unitary and upper triangular implies that

$$R_N = \text{diag}(\cdots e^{i\theta_j} \cdots),$$

for some θ_j, $j = 1, \ldots, N$. Without loss of generality we may assume that $R_N = I_N$. The upper Hessenberg structure of H_N allows for the QR factorization to be computed efficiently using elementary unitary matrices

$$H_N = G_1 G_2 \cdots G_{N-1} R_N, \tag{5}$$

where

$$G_j = \begin{bmatrix} I_{j-1} & & \\ & -\gamma_j & \sigma_j \\ & \sigma_j & \bar{\gamma}_j \\ & & & I_{N-j-1} \end{bmatrix},$$

and $\gamma_j \in \mathcal{C}$, $|\gamma_j| \leq 1$, and $\sigma_j = (1 - |\gamma_j|^2)^{1/2}$. Since each G_j is a rank 2 correction of the identity matrix, it follows that the information content of H_N is of order N. By substituting (5) into (4) and equating columns we obtain:

$$U\underline{v}_0 = -\gamma_1 \underline{v}_0 + \sigma_1 \underline{v}_1$$

$$\gamma_1 = -\langle U\underline{v}_0, \underline{v}_0 \rangle$$

$$\sigma_1 = (1 - |\gamma_1|^2)^{1/2}$$

$$U\underline{v}_1 = -\gamma_2(\sigma_1 \underline{v}_0 + \bar{\gamma}_1 \underline{v}_1) + \sigma_2 \underline{v}_2$$

$$\gamma_2 = -\langle U\underline{v}_1, \sigma_1 \underline{v}_0 + \bar{\gamma}_1 \underline{v}_1 \rangle$$

$$\sigma_2 = (1 - |\gamma_2|^2)^{1/2}$$

$$\vdots$$

Rearranging yields the Gragg algorithm:

$$\underline{v}_0 = \tilde{\underline{v}}_0 = \frac{\underline{r}_0}{\|\underline{r}_0\|}$$

$$\vdots$$

$$\gamma_i = -\langle U\underline{v}_{i-1}, \tilde{\underline{v}}_{i-1} \rangle$$

$$\sigma_i = (1 - |\gamma_i|^2)^{1/2}$$

$$\sigma_i \underline{v}_i = U\underline{v}_{i-1} + \gamma_i \tilde{\underline{v}}_{i-1}$$

$$\tilde{\underline{v}}_i = (\sigma_i \tilde{\underline{v}}_{i-1} + \bar{\gamma}_i \underline{v}_i)$$

$$\vdots$$

See ([4],[6]) for a more complete presentation. Another derivation of this double recursion for unitary matrices is given by Watkins [10].

2.2 A minimal residual algorithm for shifted unitary matrices

Jagels and Reichel [6] use the Gragg algorithm to construct an efficient minimal residual algorithm for the solution of the linear system, $A\underline{x} = \underline{b}$, where A is a shifted unitary matrix

(2). They observe that $\mathcal{K}_\ell(\underline{r}_0, A) = \mathcal{K}_\ell(\underline{r}_0, U)$. Thus, the Gragg algorithm can be used to construct an orthonormal basis for $\mathcal{K}_\ell(\underline{r}_0, A)$. This basis is then used to construct a minimal residual algorithm as follows:

At step ℓ of the Gram-Schmidt process, the recursions for the basis vectors, $\{\underline{v}_j\}_{j=0}^\ell$, can be written in matrix form as

$$(6) \qquad UV_\ell = V_{\ell+1} H^{(U)}_{\ell+1,\ell},$$

where $H^{(U)}_{\ell+1,\ell}$ is the $(\ell+1) \times \ell$ upper Hessenberg matrix,

$$H^{(U)}_{\ell+1,\ell} = G_1 G_2 \cdots G_{\ell-1} \tilde{G}_\ell,$$

and

$$G_j = \begin{bmatrix} I_{j-1} & & \\ & -\gamma_j & \sigma_j \\ & \sigma_j & \bar{\gamma}_j \\ & & & I_{\ell-j-1} \end{bmatrix}, \quad \tilde{G}_\ell = \begin{bmatrix} I_{\ell-1} & \\ & \gamma_\ell \\ & \sigma_\ell \end{bmatrix}.$$

Using (2) and (6) we see that

$$(7) \qquad AV_\ell = (\rho U + \xi I)V_\ell = V_{\ell+1}(\rho H^{(U)}_{\ell+1,\ell} + \xi \hat{I}_\ell) = V_{\ell+1} H^{(A)}_{\ell+1,\ell},$$

where

$$\hat{I}_\ell = \begin{bmatrix} I_\ell \\ \underline{0}^T \end{bmatrix} \in \mathcal{C}^{(\ell+1)\times \ell}.$$

We remark here, that instead of deriving $H^{(A)}_{\ell+1,\ell}$ from $H^{(U)}_{\ell+1,\ell}$, one could obtain $H^{(A)}_{\ell+1,\ell}$ directly by substituting $U = \frac{1}{\rho}(A - \xi I)$ into the Gragg algorithm. Now, the standard argument for constructing a GMRES algorithm (c.f.[9]) is employed. Since $\|\underline{r}_0\| \underline{v}_0 = \underline{r}_0$, we have for some $\underline{y}_\ell \in \mathcal{C}^\ell$

$$\underline{r}_\ell = \underline{r}_0 - AV_\ell \underline{y}_\ell = V_{\ell+1}[\|\underline{r}_0\|\epsilon_1 - H^{(A)}_{\ell+1,\ell} \underline{y}_\ell],$$

where $\epsilon_1 = [1, 0, ..., 0]^T$. The objective is to choose $\underline{y}_\ell$ to minimize $\|\underline{r}_\ell\|$. Since $V_{\ell+1}$ has orthonormal columns, this may be accomplished by solving the $(\ell+1) \times \ell$ least squares problem

$$\|\underline{r}_0\|\epsilon_1 \approx H^{(A)}_{\ell+1,\ell} \underline{y}_\ell.$$

Since $(\rho H^{(U)}_{\ell+1,\ell} + \xi \hat{I}_\ell) = H^{(A)}_{\ell+1,\ell}$ is upper Hessenberg, a QR factorization of $H^{(A)}_{\ell+1,\ell}$ is accomplished by Givens matrices. The relationship between $H^{(A)}_{\ell+1,\ell}$ and $H^{(U)}_{\ell+1,\ell}$ allows the least squares problem to be solved using an algorithm that involves five short recursions. See ([6]) for details.

3 B-unitary and shifted B-unitary matrices

3.1 A new minimal residual algorithm for shifted unitary matrices

In this section, we present an alternative minimal residual algorithm for shifted unitary matrices. This reformulation will motivate the development in Section 4. First, let us restate the minimal residual method for the solution of the linear system, $A\underline{x} = \underline{b}$, as a conjugate gradient method whose iterates are uniquely defined by:

$$\begin{aligned} \underline{x}_{i+1} &= \underline{x}_i + \alpha_i \underline{p}_i, \quad \underline{p}_i \in \mathcal{K}_{i+1}(\underline{r}_0, A), \\ \underline{e}_{i+1} &\perp_{A^*A} \mathcal{K}_{i+1}(\underline{r}_0, A). \end{aligned}$$

In this context, it is clear that the direction vectors, $\underline{p}_i$'s, must satisfy

$$\underline{p}_i \in \mathcal{K}_{i+1}(\underline{r}_0, A),$$
$$\underline{p}_i \perp_{A^*A} \mathcal{K}_i(\underline{r}_0, A).$$

Suppose that $d(\underline{r}_0, A) = N$. We seek a basis, $\{\underline{p}_0, \underline{p}_1, ..., \underline{p}_{N-1}\}$, for $\mathcal{K}_N(\underline{r}_0, A)$, such that

$$\langle A^*A\underline{p}_i, \underline{p}_j\rangle = \begin{cases} 0 & i \neq j, \\ 1 & i = j. \end{cases}$$

Now, suppose that A is shifted unitary, i.e., $A = \rho U + \xi I$, where U is unitary and that $N = d(\underline{r}_0, U) = d(\underline{r}_0, A)$. Note that in this case $\mathcal{K}_\ell(\underline{r}_0, A) = \mathcal{K}_\ell(\underline{r}_0, U)$. Thus, it is sufficient to find an A^*A orthogonal basis for $\mathcal{K}_\ell(\underline{r}_0, U)$. Next, note that if U is unitary with respect to I, and $A = \rho U + \xi I$, then

(8) $$A^*AU = UA^*A.$$

It follows that U is A^*A-unitary, or in other words, U is isometric with respect to the inner product $\langle A^*A\cdot, \cdot\rangle$,

(9) $$\langle A^*AU\underline{x}, U\underline{y}\rangle = \langle UA^*A\underline{x}, U\underline{y}\rangle = \langle A^*A\underline{x}, \underline{y}\rangle, \quad \forall \underline{x}, \underline{y} \in \mathcal{C}^N.$$

Gragg [4] gave a more general result than what we presented in Section 2.1. This result says that if a matrix U_B is isometric with respect to any inner product, $\langle B\cdot, \cdot\rangle$, then a B-orthonormal basis for $\mathcal{K}_\ell(\underline{r}_0, U_B)$ can be constructed with a short double recursion of the form given in Section 2.1. We show how this is done in the shifted unitary case, where U is isometric with respect to the inner product $\langle A^*A\cdot, \cdot\rangle$. Consider a Gram-Schmidt process to construct an A^*A-orthonormal basis for $\mathcal{K}_N(\underline{r}_0, U)$. At step N, the recursions can be written in matrix form as

(10) $$UP_N = P_N H_N,$$

where H_N is a $N \times N$ upper Hessenberg matrix. Multiplying both sides of (10) by $P_N^*U^*A^*A$, we obtain

$$P_N^*U^*A^*AUP_N = P_N^*U^*A^*AP_N H_N.$$

Using (8) and (10) this becomes

$$P_N^*A^*AP_N = H_N^*P_N^*A^*AP_N H_N.$$

Since P_N is constructed so that $P_N^*A^*AP_N = I_N$, it follows that $I_N = H_N^*H_N$, and H_N is I-unitary. Therefore, H_N can be written as a product of Givens matrices. In fact, the iteration to produce an A^*A-orthonormal basis is exactly like the Gragg algorithm except that the standard inner product $\langle \cdot, \cdot\rangle$ is replaced by $\langle A^*A\cdot, \cdot\rangle$. The algorithm is:

$$\underline{p}_0 = \underline{\tilde{p}}_0 = \underline{r}_0/\|\underline{r}_0\|_{A^*A},$$
$$\vdots$$
$$\gamma_i = -\langle AU\underline{p}_{i-1}, A\underline{\tilde{p}}_{i-1}\rangle,$$
$$\sigma_i = (1 - |\gamma_i|^2)^{1/2},$$
$$\sigma_i \underline{p}_i = U\underline{p}_{i-1} + \gamma_i \underline{\tilde{p}}_{i-1}$$
$$\underline{\tilde{p}}_i = (\sigma_i \underline{\tilde{p}}_{i-1} + \bar{\gamma}_i \underline{p}_i),$$
$$\vdots$$

Given $\underline{x}_0$, the basis thus computed can be used to solve the linear system, $A\underline{x} = \underline{b}$, by adding at each step the recursions,

$$\underline{x}_{i+1} = \underline{x}_i + \alpha_i \underline{p}_i, \quad \alpha_i = \frac{\langle A^* A \underline{e}_i, \underline{p}_i \rangle}{\langle A^* A \underline{p}_i, \underline{p}_i \rangle} = \langle \underline{r}_i, A\underline{p}_i \rangle,$$
$$\underline{r}_{i+1} = \underline{r}_i - \alpha_i A \underline{p}_i.$$

Note that the algorithm above requires multiplications by both U and A. To rearrange the algorithm to one requiring only multiplication by U, define the quantities

$$\underline{q}_i = A\underline{p}_i \quad \text{and} \quad \underline{\tilde{q}}_i = A\underline{\tilde{p}}_i.$$

Notice that for $A = \rho U + \xi I$, A and U commute, thus, multiplying the equations for $\underline{p}_i$ and $\underline{\tilde{p}}_i$ by A, we obtain

$$\sigma_i \underline{q}_i = U \underline{q}_{i-1} + \gamma_i \underline{\tilde{q}}_{i-1}, \quad \text{and} \quad \underline{\tilde{q}}_i = (\sigma_i \underline{\tilde{q}}_{i-1} + \bar{\gamma}_i \underline{q}_i).$$

Also, since $A\underline{p}_i = \rho U \underline{p}_i + \xi \underline{p}_i$, we can write

$$U\underline{p}_i = \frac{1}{\rho}(\underline{q}_i - \xi \underline{p}_i).$$

Using this information we rewrite the above minimal residual algorithm for shifted unitary matrices.

$$\underline{p}_0 = \underline{\tilde{p}}_0 = \underline{r}_0/\|\underline{r}_0\|_{A^*A},$$
$$\underline{q}_0 = \underline{\tilde{q}}_0 = A\underline{p}_0,$$
$$\underline{x}_1 = \underline{x}_0 + \alpha_0 \underline{p}_0, \quad \alpha_0 = \langle \underline{r}_0, \underline{q}_0 \rangle,$$
$$\underline{r}_1 = \underline{r}_0 - \alpha_0 \underline{q}_0,$$
$$\vdots$$
$$\gamma_i = -\langle U\underline{q}_{i-1}, \underline{\tilde{q}}_{i-1} \rangle,$$
$$\sigma_i = (1 - |\gamma_i|^2)^{1/2},$$
$$\underline{q}_i = \frac{1}{\sigma_i}(U\underline{q}_{i-1} + \gamma_i \underline{\tilde{q}}_{i-1}),$$
$$\underline{p}_i = \frac{1}{\rho \sigma_i}(\underline{q}_{i-1} - \xi \underline{p}_{i-1}) + \frac{\gamma_i}{\sigma_i} \underline{\tilde{p}}_{i-1},$$
$$\underline{\tilde{q}}_i = \sigma_i \underline{\tilde{q}}_{i-1} + \bar{\gamma}_i \underline{q}_i,$$
$$\underline{\tilde{p}}_i = \sigma_i \underline{\tilde{p}}_{i-1} + \bar{\gamma}_i \underline{p}_i,$$
$$\underline{x}_{i+1} = \underline{x}_i + \alpha_i \underline{p}_i, \quad \alpha_i = \langle \underline{r}_i, \underline{q}_i \rangle,$$
$$\underline{r}_{i+1} = \underline{r}_i - \alpha_i \underline{q}_i,$$
$$\vdots$$

This algorithm requires the storage of 7 vectors:
$$\underline{x}_i,\ \underline{r}_i,\ \underline{p}_{i-1},\ \underline{\tilde{p}}_{i-1},\ \underline{q}_{i-1},\ \underline{\tilde{q}}_{i-1},\ U\underline{q}_{i-1}.$$

The approximate cost per iteration is:
1). 1 Matrix vector multiplication: $U\underline{q}_{i-1}$
2). 2 inner products: $\langle \underline{r}_i, \underline{q}_i \rangle$, $\langle U\underline{q}_{i-1}, \underline{\tilde{q}}_{i-1}\rangle$
3). 7 SAXPY.

The computational and storage requirements are comparable to the Jagels and Reichel algorithm, which requires approximately 1 matrix vector multiplication, 1 inner product, 6 SAXPY, and the storage of 6 vectors.

3.2 CG algorithms for general B-unitary and shifted B-unitary matrices

For shifted unitary matrices (2), U is isometric with respect to the inner product $\langle A^*A\cdot,\cdot\rangle$. Thus, an A^*A orthonormal basis for $\mathcal{K}_N(\underline{r}_0, A)$ can be generated with a pair of short recursions. Consequently, a minimal residual method, $CG(B, A)$, with $B = A^*A$, can be implemented practically.

These results can be generalized to obtain practical conjugate gradient algorithms for any matrices that are isometric with respect to some inner product $\langle B\cdot,\cdot\rangle$ and for which $\langle B\underline{e}_i, \underline{p}_i\rangle$ is computable (see below). We refer to these matrices as B-unitary, and denote them by U_B. They satisfy

$$\langle BU_B\underline{x}, U_B\underline{y}\rangle = \langle B\underline{x}, \underline{y}\rangle, \quad \forall \underline{x},\ \underline{y} \in \mathcal{C}^N.$$

Matrix notation after step N of a Gram-Schmidt process to construct a B-orthonormal basis for $\mathcal{K}_N(\underline{r}_0, U_B)$ is given by
$$U_B P_N = P_N H_N, \tag{11}$$

where H_N is a $N \times N$ upper Hessenberg matrix. Multiplying through on the left by $P_N^* U_B^* B$ and then substituting in $P_N^* U_B^* = H_N^* P_N^*$ and $U_B^* B U_B = B$ in (11) yields

$$P_N^* B P_N = H_N^* P_N^* B P_N H_N.$$

Since $P_N^* B P_N = I_N$ by construction, it follows that H_N is I-unitary, thus it can be factored into a product of elementary unitary matrices. Analogous to the unitary case in Section 2, the substitution of this factorization for H_N into (11) and equating the columns yields the recursion formulas for the basis vectors.

A shifted B-unitary matrix A_B has the form
$$A_B = \rho U_B + \xi I.$$

Since $\mathcal{K}_N(\underline{r}_0, A_B) = \mathcal{K}_N(\underline{r}_0, U_B)$, the same B-orthonormal basis for $\mathcal{K}_N(\underline{r}_0, U_B)$ can be used for $\mathcal{K}_N(\underline{r}_0, A_B)$. Thus a conjugate gradient method for shifted B-unitary matrices can also be implemented with short recursions.

An algorithm for $CG(B, A)$, where A is either B-unitary, or shifted B-unitary, can be obtained as follows: Construct a B-orthonormal basis for $\mathcal{K}_N(\underline{r}_0, A)$ using the Gragg algorithm generalized to the B-inner product. At each step i, perform the additional recursions,

$$\underline{x}_{i+1} = \underline{x}_i + \alpha_i \underline{p}_i, \quad \alpha_i = \frac{\langle B\underline{e}_i, \underline{p}_i\rangle}{\langle B\underline{p}_i, \underline{p}_i\rangle} = \langle B\underline{e}_i, \underline{p}_i\rangle,$$
$$\underline{r}_{i+1} = \underline{r}_i - \alpha_i A\underline{p}_i.$$

Note that, since the computation of the α_i's involves the unknown quantity of the error, B must be chosen so that α_i is computable.

4 Single (ℓ, m) recursions
4.1 CG using single (ℓ, m) recursions

In Section 3, we saw that if A is a shifted B-unitary matrix, i.e., $A = A_B = \rho U_B + \xi I$, a double recursion,

$$\begin{aligned}(12)\qquad \sigma_i \underline{p}_i &= U_B \underline{p}_{i-1} + \gamma_i \underline{\tilde{p}}_{i-1}, \\ \underline{\tilde{p}}_i &= \sigma_i \underline{\tilde{p}}_{i-1} + \bar{\gamma}_i \underline{p}_i,\end{aligned}$$

can be used to construct a B-orthonormal basis for $\mathcal{K}_N(\underline{r}_0, A)$, where $N = d(\underline{r}_0, A)$. At step $i + 1$, form

$$\sigma_{i+1} \underline{p}_{i+1} = U_B \underline{p}_i + \gamma_{i+1} \underline{\tilde{p}}_i.$$

For $\underline{\tilde{p}}_i$, substitute the second equation in (12) to obtain

$$\sigma_{i+1} \underline{p}_{i+1} = U_B \underline{p}_i + \gamma_{i+1}(\sigma_i \underline{\tilde{p}}_{i-1} + \bar{\gamma}_i \underline{p}_i).$$

Solving the first equation in (12) for $\underline{\tilde{p}}_{i-1}$, and making this substitution into the above yields the recursion,

$$(13)\qquad \underline{p}_{i+1} = \frac{1}{\sigma_{i+1}} U_B \underline{p}_i - \frac{\gamma_{i+1}\sigma_i}{\gamma_i \sigma_{i+1}} U_B \underline{p}_{i-1} + \left(\frac{\gamma_{i+1}\sigma_i^2}{\gamma_i \sigma_{i+1}} + \frac{\gamma_{i+1}\bar{\gamma}_i}{\sigma_{i+1}}\right)\underline{p}_i.$$

With the substitution of $U_B = \frac{1}{\rho}(A - \xi I)$, (13) becomes

$$(14)\qquad \begin{aligned}\underline{p}_{i+1} &= \frac{1}{\rho\sigma_{i+1}} A\underline{p}_i - \frac{\gamma_{i+1}\sigma_i}{\rho\gamma_i\sigma_{i+1}} A\underline{p}_{i-1} \\ &+ \left(\frac{\gamma_{i+1}\sigma_i^2}{\gamma_i\sigma_{i+1}} - \frac{\xi}{\rho\sigma_{i+1}} + \frac{\gamma_{i+1}\bar{\gamma}_i}{\sigma_{i+1}}\right)\underline{p}_i + \frac{\gamma_{i+1}\sigma_i\xi}{\rho\gamma_i\sigma_{i+1}}\underline{p}_{i-1}.\end{aligned}$$

Thus, for shifted B-unitary matrices, the double recursion (12) can be rewritten as a single recursion (14), involving the two previous $A\underline{p}$'s and the two previous $\underline{p}$'s. (Notice that, if A is B-unitary, i.e., $A = U_B$, the double recursion (12) still yields a B-orthonormal basis for $\mathcal{K}_N(\underline{r}_0, A)$, and it can be rewritten as the single recursion given by (13).)

In this section, we will consider general recursions of the form

$$(15)\qquad \underline{p}_{i+1} = \sum_{j=i-m}^{i} \beta_{i,j} A\underline{p}_j - \sum_{j=i-\ell}^{i} \sigma_{i,j} \underline{p}_j,$$

where ℓ and m can be any integers ≥ 0. If a recursion of this form yields a B-orthonormal basis for $\mathcal{K}_N(\underline{r}_0, A)$, for some integers $\ell, m \geq 0$, we say $A \in CG_B(\ell, m)$. Note that, $\underline{p}_{i+1}$ can always be computed by a recursion of the form

$$\underline{p}_{i+1} = \sum_{j=i-m}^{i} \beta_{i,j} A\underline{p}_j - \sum_{j=0}^{i} \sigma_{i,j} \underline{p}_j,$$

because, given any $\{\beta_{i,j}\}_{j=i-m}^{i}$, there exists $\{\sigma_{i,j}\}_{j=0}^{i}$ given by

$$\sigma_{i,j} = \frac{\langle \sum_{j=i-m}^{i} \beta_{i,j} A\underline{p}_j, \underline{p}_j\rangle_B}{\langle \underline{p}_j, \underline{p}_j \rangle_B}, \qquad j = 0, ..., i.$$

However, we are interested in determining the class of matrices for which a B-orthonormal basis can be constructed using a short recursion of the form (15), that is, with both ℓ and m small. With this in mind, we make the following definition.

DEFINITION 4.1. $A \in \mathcal{CG}_B(\ell, m)$ if for every $\underline{p}_0$ and every $0 \leq i \leq d(\underline{p}_0, A) - 1$, there exists coefficients $\{\beta_{i,j}\}_{j=i-m}^{i}$ such that

$$(16) \qquad \sigma_{i,k} = \frac{\langle \sum_{j=i-m}^{i} \beta_{i,j} A \underline{p}_j, \underline{p}_k \rangle_B}{\langle \underline{p}_k, \underline{p}_k \rangle_B} = 0, \qquad \text{for } k = 0, ..., i - \ell - 1.$$

We note here, that the recursion (15) is of a more general form than that given by (1). By letting $m = 0$ and $\ell = s - 2$ in (15) we obtain the form (1). Also, the direction vectors for unitary and shifted unitary matrices can be constructed with a short (ℓ, m) recursion, whereas they cannot be obtained with a short recursion of the form (1). In a similar manner as in the unitary and shifted unitary cases, it can be shown that many short multiple recursions can be rewritten as single (ℓ, m) recursions, for some small ℓ, m. These multiple recursions can be analyzed as single (ℓ, m) recursions.

4.2 Sufficient conditions for $A \in \mathcal{CG}_B(\ell, m)$

We first review some terminology on normal matrices. Recall that a conjugate gradient method is defined with respect to an HPD inner product matrix B, and the system matrix A. The B-adjoint of A is the unique matrix, $A^\dagger$, satisfying

$$\langle BA\underline{x}, \underline{y} \rangle = \langle B\underline{x}, A^\dagger \underline{y} \rangle, \qquad \forall \underline{x}, \underline{y} \in \mathcal{C}^N.$$

This yields, $A^\dagger = B^{-1} A^* B$. The matrix A is B-normal if it satisfies the following equivalent conditions:

i) $AA^\dagger = A^\dagger A$,

ii) $A, A^\dagger$ have the same complete set of B-orthogonal eigenvectors,

iii) $A^\dagger$ can be written as a polynomial of some degree η in the matrix A.

We say that A is B-normal(s) if it is B-normal, and s is the smallest degree for which $A^\dagger = P_\eta(A)$.

DEFINITION 4.2. A is B-normal(ℓ, m) if A is B-normal and there exists polynomials $p_\ell(\lambda)$ and $q_m(\lambda)$, of degree ℓ and m respectively, such that

$$(17) \qquad A^\dagger q_m(A) = p_\ell(A).$$

Since $A^\dagger = \frac{p_\ell(A)}{q_m(A)}$, we say A is B-normal with rational degree ℓ/m.

Note that if $m = 0$, (17) becomes

$$A^\dagger = \hat{p}_\ell(A),$$

and A is B-normal(ℓ).

The following Lemma gives sufficient conditions on the matrix A for $A \in \mathcal{CG}_B(\ell, m)$.

LEMMA 4.1. If A is B-normal(ℓ, m), then $A \in \mathcal{CG}_B(\ell, m)$.

Proof. In order for $A \in \mathcal{CG}_B(\ell, m)$, (16) must be satisfied, or equivalently,

$$\sigma_{i,k} = \langle \sum_{j=i-m}^{i} \beta_{i,j}\underline{p}_j, A^\dagger \underline{p}_k \rangle_B = 0, \qquad k = 0, ..., i - \ell - 1.$$

There are polynomials, ψ_k, $\tilde{\psi}_{k-m}$, and $r_{m-1}^{(k)}$ such that

$$\underline{p}_k = \psi_k(A)\underline{p}_0 = q_m(A)\tilde{\psi}_{k-m}(A)\underline{p}_0 + r_{m-1}^{(k)}(A)\underline{p}_0,$$

where the second equality follows upon division of $\psi_k(\lambda)$ by $q_m(\lambda)$. After multiplying by $A^\dagger$, since A is B-normal(ℓ, m), we can use (17) to yield

$$\begin{aligned} A^\dagger \underline{p}_k &= A^\dagger q_m(A)\tilde{\psi}_{k-m}(A)\underline{p}_0 + A^\dagger r_{m-1}^{(k)}(A)\underline{p}_0 \\ &= p_\ell(A)\tilde{\psi}_{k-m}(A)\underline{p}_0 + A^\dagger r_{m-1}^{(k)}(A)\underline{p}_0. \end{aligned}$$

For $k = 0, ..., i - \ell - 1$,

$$\begin{aligned} p_\ell(A)\tilde{\psi}_{k-m}(A)\underline{p}_0 &\in \text{sp}\{\underline{p}_0, \underline{p}_1, ..., \underline{p}_{i-m-1}\}, \\ &\text{and} \\ A^\dagger r_{m-1}^{(k)}(A)\underline{p}_0 &\in \text{sp}\{A^\dagger \underline{p}_0, A^\dagger \underline{p}_1, ..., A^\dagger \underline{p}_{m-1}\}. \end{aligned}$$

Since $\text{sp}\{\underline{p}_0, \underline{p}_1, ..., \underline{p}_{i-m-1}\}$ is B-orthogonal to $\text{sp}\{\underline{p}_{i-m}, ..., \underline{p}_i\}$, we only need to choose $\{\beta_{i,j}\}_{j=i-m}^{i}$ so that

$$\langle \sum_{j=i-m}^{i} \beta_{i,j}\underline{p}_j, A^\dagger r_{m-1}^{(k)}(A)\underline{p}_0 \rangle_B = 0, \qquad \text{for } k = 0, ..., i - \ell - 1,$$

or equivalently, choose $\{\beta_{i,j}\}_{j=i-m}^{i}$ to satisfy

$$(18) \quad \begin{bmatrix} \langle \underline{p}_i, A^\dagger \underline{p}_0 \rangle_B & \langle \underline{p}_{i-1}, A^\dagger \underline{p}_0 \rangle_B & \cdots & \langle \underline{p}_{i-m}, A^\dagger \underline{p}_0 \rangle_B \\ \langle \underline{p}_i, A^\dagger \underline{p}_1 \rangle_B & \langle \underline{p}_{i-1}, A^\dagger \underline{p}_1 \rangle_B & \cdots & \langle \underline{p}_{i-m}, A^\dagger \underline{p}_1 \rangle_B \\ \vdots & \vdots & & \vdots \\ \langle \underline{p}_i, A^\dagger \underline{p}_{m-1} \rangle_B & \langle p_{i-1}, A^\dagger \underline{p}_{m-1} \rangle_B & \cdots & \langle \underline{p}_{i-m}, A^\dagger \underline{p}_{m-1} \rangle_B \end{bmatrix} \begin{pmatrix} \beta_{i,i} \\ \beta_{i,i-1} \\ \vdots \\ \beta_{i,i-m} \end{pmatrix} = \begin{pmatrix} 0 \\ 0 \\ \vdots \\ 0 \end{pmatrix}.$$

This system can always be solved for the $\beta_{i,j}$'s. Therefore, if A is B-normal(ℓ, m), then $A \in \mathcal{CG}_B(\ell, m)$. $\square$

REMARK 4.1. *While the system (18) always has a nontrivial solution, unless a solution exists for which $\beta_{i,i} \neq 0$, the desired result will not be achieved. If $\beta_{i,i} = 0$, equation (15) will yield $\underline{p}_{i+1} \in \mathcal{K}_{i-1}(\underline{p}_0, A)$ such that $\underline{p}_{i+1} \perp \mathcal{K}_{i-1}(\underline{p}_0, A)$, that is, $\underline{p}_{i+1} \equiv \underline{0}$. We call this condition breakdown of the algorithm. Using techniques similar to those in ([8]) it can be shown that breakdown occurrs for $\underline{p}_0$ in a set of measure zero. Moreover, there is a reorganization of the computations that avoids breakdown. Details will appear in [2].*

Next we characterize B-normal(ℓ, m) matrices. Recall that B-normal(ℓ, m) matrices satisfy

$$(19) \qquad A^\dagger = \frac{p_\ell(A)}{q_m(A)} = \frac{a_\ell A^\ell + \cdots + a_1 A + a_0 I}{b_m A^m + \cdots + b_1 A + b_0 I},$$

for some polynomials,

$$p_\ell(\mu) = a_\ell \mu^\ell + \cdots + a_1 \mu + a_0 \quad \text{and} \quad q_m(\mu) = b_m \mu^m + \cdots + b_1 + b_0,$$

of degree ℓ and m respectively. In the following Theorem which characterizes B-normal(ℓ, m) matrices, we will assume that ℓ and m are the smallest degrees for which (19) holds. This means that $a_\ell, b_m \neq 0$, and that $p_\ell(\mu)$ and $q_m(\mu)$ have no common roots, since otherwise, (19) would hold for some smaller degrees ℓ' and m'.

THEOREM 4.1. *Let A be B-normal(ℓ, m), where ℓ and m are the smallest degrees for which*

$$A^\dagger = \frac{p_\ell(A)}{q_m(A)}.$$

Let $d(A)$ be the degree of the minimal polynomial of A. Then,

1) *If $\ell > m + 1$, then $d(A) \leq \ell^2$,*

2) *if $\ell = m + 1$, then $d(A) \leq \ell^2$ or A is B-normal$(1, 0)$,*

3) *if $\ell < m$, and $b_0 \neq 0$, then $d(A) \leq m^2 + 1$,*

4) *if $\ell < m - 1$ and $b_0 = 0$, then $d(A) \leq m^2$,*

5) *if $\ell = m - 1$ and $b_0 = 0$, then $d(A) \leq m^2$, or A is B-normal$(0, 1)$,*

6) *if $\ell = m$, then $d(A) \leq m^2 + 1$, or A is B-normal$(1, 1)$,*

Proof. From (19) it follows that the eigenvalues of a B-normal(ℓ, m) matrix can be written as:

(20) $$\bar{\lambda}_i = \frac{p_\ell(\lambda_i)}{q_m(\lambda_i)}, \quad \forall \lambda_i \in \Sigma(A).$$

We assume that $p_\ell(\lambda_i) \neq 0$, $\forall \lambda_i \in \Sigma(A)$, since otherwise, (20) would imply that $\lambda_i = 0$ and A is singular. We further assume that $q_m(\lambda_i) \neq 0$, $\forall \lambda_i \in \Sigma(A)$, since $q_m(\lambda_i) = 0$ and $p_\ell(\lambda_i) \neq 0$ for some $\lambda_i \in \Sigma(A)$ would imply that $\lambda_i = \infty$. Taking conjugates in (20) yields

$$\lambda_i = \frac{\bar{p}_\ell(\bar{\lambda}_i)}{\bar{q}_m(\bar{\lambda}_i)} = \frac{\bar{p}_\ell\left(\frac{p_\ell(\lambda_i)}{q_m(\lambda_i)}\right)}{\bar{q}_m\left(\frac{p_\ell(\lambda_i)}{q_m(\lambda_i)}\right)}, \quad \forall \lambda_i \in \Sigma(A).$$

This can be expanded into

$$\lambda_i = \frac{\bar{a}_\ell \left(\frac{p_\ell(\lambda_i)}{q_m(\lambda_i)}\right)^\ell + \bar{a}_{\ell-1}\left(\frac{p_\ell(\lambda_i)}{q_m(\lambda_i)}\right)^{\ell-1} + \cdots + \bar{a}_0}{\bar{b}_m\left(\frac{p_\ell(\lambda_i)}{q_m(\lambda_i)}\right)^m + \bar{b}_{m-1}\left(\frac{p_\ell(\lambda_i)}{q_m(\lambda_i)}\right)^{m-1} + \cdots + \bar{b}_0}$$

$$= \frac{(q_m(\lambda_i))^m}{(q_m(\lambda_i))^\ell} \cdot \frac{\bar{a}_\ell(p_\ell(\lambda_i))^\ell + \bar{a}_{\ell-1}(p_\ell(\lambda_i))^{\ell-1} q_m(\lambda_i) + \cdots + \bar{a}_0(q_m(\lambda_i))^\ell}{\bar{b}_m(p_\ell(\lambda_i))^m + \bar{b}_{m-1}(p_\ell(\lambda_i))^{m-1} q_m(\lambda_i) + \cdots + \bar{b}_0(q_m(\lambda_i))^m},$$

$\forall \lambda_i \in \Sigma(A)$. Cross multiplication and collection of terms on one side yields

(21) $$\lambda_i(q_m(\lambda_i))^\ell \left[\bar{b}_m(p_\ell(\lambda_i))^m + \bar{b}_{m-1}(p_\ell(\lambda_i))^{m-1}q_m(\lambda_i) + \cdots + \bar{b}_0(q_m(\lambda_i))^m\right] \\ -(q_m(\lambda_i))^m \left[\bar{a}_\ell(p_\ell(\lambda_i))^\ell + \bar{a}_{\ell-1}(p_\ell(\lambda_i))^{\ell-1}q_m(\lambda_i) + \cdots + \bar{a}_0(q_m(\lambda_i))^\ell\right] = 0,$$

$\forall \lambda_i \in \Sigma(A)$.

First, consider the case when A is B-normal(ℓ, m) with $\ell > m$. Since $q_m(\lambda_i) \neq 0$, $\forall \lambda_i \in \Sigma(A)$, we can divide (21) by $(q_m(\lambda_i))^m$. It follows that the polynomial

(22) $$\mu(q_m(\mu))^{\ell-m} \left[\bar{b}_m(p_\ell(\mu))^m + \bar{b}_{m-1}(p_\ell(\mu))^{m-1}q_m(\mu) + \cdots + \bar{b}_0(q_m(\mu))^m\right] \\ -\bar{a}_\ell(p_\ell(\mu))^\ell - \bar{a}_{\ell-1}(p_\ell(\mu))^{\ell-1}q_m(\mu) - \cdots - \bar{a}_0(q_m(\mu))^\ell = 0,$$

whenever $\mu = \lambda_i$, and $\lambda_i \in \Sigma(A)$.

Proof of 1): If $\ell > m+1$ and $m = 0$, A is B-normal(ℓ). In [3] it was shown that these matrices have less that or equal to ℓ^2 distinct eigenvalues. Therefore, $d(A) \leq \ell^2$.

If $\ell > m+1$ and $m > 0$, the highest degree term of the polynomial in (22) is

$$-\bar{a}_\ell(a_\ell\mu^\ell)^\ell,$$

with degree ℓ^2. By hypothesis, $a_\ell \neq 0$, which implies (22) has at most ℓ^2 distinct roots. Thus, $d(A) \leq \ell^2$.

Proof of 2): If $\ell = m+1$ and $m = 0$, A is B-normal(1). It was shown in [3] that the eigenvalues of these matrices are collinear, that is, they lie on a straight line in the complex plane.

If $\ell = m+1$ and $m > 0$, the highest degree term of the polynomial in (22) is

$$\mu(b_m\mu^m)\bar{b}_m(a_\ell\mu^\ell)^m - \bar{a}_\ell(a_\ell\mu^\ell)^\ell,$$

with degree ℓ^2. Either the polynomial has at most ℓ^2 distinct roots, or (22) holds for all μ. In particular, (22) holds for the roots of $q_m(\mu)$, say μ_j, for $j = 1, ..., m$. Plugging μ_j into (22) yields

$$\bar{a}_\ell(p_\ell(\mu_j))^\ell = 0, \quad \text{for } j = 1, ..., m.$$

This means that either $a_\ell = 0$ or $p_\ell(\mu_j) = 0$, for $j = 1, ..., m$, or both are zero. By hypothesis, $a_\ell \neq 0$ and $p_\ell(\mu)$ and $q_m(\mu)$ have no common roots, so it follows that (22) cannot hold for all μ, thus $d(A) \leq \ell^2$.

Next, consider the case when A is B-normal(ℓ, m), with $\ell \leq m$. Since $q_m(\lambda_i) \neq 0$, $\forall \lambda_i \in \Sigma(A)$, we can divide (21) by $(q_m(\lambda_i))^\ell$. We obtain the polynomial

(23) $$\mu \left[\bar{b}_m(p_\ell(\mu))^m + \bar{b}_{m-1}(p_\ell(\mu))^{m-1}q_m(\mu) + \cdots + \bar{b}_0(q_m(\mu))^m\right] \\ -(q_m(\mu))^{m-\ell} \left[\bar{a}_\ell(p_\ell(\mu))^\ell + \bar{a}_{\ell-1}(p_\ell(\mu))^{\ell-1}q_m(\mu) + \cdots + \bar{a}_0(q_m(\mu))^\ell\right] = 0,$$

whenever $\mu = \lambda_i$, and $\lambda_i \in \Sigma(A)$. Since $p_\ell(\lambda)$ and $q_m(\lambda)$ are polynomials of degree ℓ and m respectively, a_ℓ and b_m are nonzero. However, it is possible that any of the other coefficients, $a_0, ..., a_{\ell-1}$ and $b_0, ..., b_{m-1}$, could be zero.

Proof of 3): If $\ell < m$ and $b_0 \neq 0$, the highest degree term of the polynomial in (23) is

$$\bar{b}_0\mu(b_m\mu^m)^m,$$

with degree $m^2 + 1$. By hypothesis, $b_0, b_m \neq 0$, and it follows that $d(A) \leq m^2 + 1$.

Proof of 4): Suppose that $\ell < m - 1$ and $b_0 = 0$. We assume that $a_0 \neq 0$, for otherwise we could could factor μ out of both $p_\ell(\mu)$ and $q_m(\mu)$, showing that A is really B-normal$(\ell - 1, m - 1)$. The highest degree term of the polynomial in (23) is

$$\bar{a}_0 (b_m \mu^m)^{m-1} (b_m \mu^m)^\ell = \bar{a}_0 (b_m \mu^m)^m,$$

with degree m^2. Either the polynomial is zero for all μ, or $d(A) \leq m^2$. By hypothesis, $a_0, b_m \neq 0$, and it follows that (23) cannot be zero for all μ, thus, $d(A) \leq m^2$.

Proof of 5): Suppose that $\ell = m - 1$ and $b_0 = 0$. We assume that $a_0 \neq 0$, since otherwise this would imply that A is B-normal$(\ell - 1, m - 1)$. There are 2 possibilities to consider, $\ell > 0$ and $\ell = 0$.

If $\ell > 0$, then $m > 1$ and it is possible for b_1 to be zero. If $b_1 = 0$, the highest degree term of the polynomial in (23) is

$$\bar{a}_0 (b_m \mu^m)^{m-1} (b_m \mu^m)^\ell = \bar{a}_0 (b_m \mu^m)^m,$$

with degree m^2. Either the polynomial has at most m^2 distinct roots, or it is zero everywhere. Since a_0 and b_m are nonzero by hypothesis, it follows that the polynomial has at most m^2 distinct roots, thus $d(A) \leq m^2$. If $b_1 \neq 0$, the highest degree term of the polynomial in (23) is

$$\bar{b}_1 \mu (b_m \mu^m)^{m-1} (a_\ell \mu^\ell) - \bar{a}_0 (b_m \mu^m)^m,$$

with degree m^2. Again, either $d(A) \leq m^2$, or (23) holds for all μ. In particular, it must hold for the roots of $p_\ell(\mu)$, say μ_j, for $j = 1, ..., \ell$. Plugging μ_j, for $j = 1, ..., \ell$ into (23) yields

$$\bar{a}_0 (q_m(\mu_j))^m = 0, \qquad j = 1, ..., \ell.$$

To satisfy (23) for all μ, requires that either $a_0 = 0$, or $q_m(\mu_j) = 0$, for $j = 1, ..., \ell$, or both are zero. By hypothesis, $a_0 \neq 0$, and $p_\ell(\mu)$ and $q_m(\mu)$ have no common roots, thus, $d(A) \leq m^2$.

If $\ell = 0$, then $m = 1$, and it follows by hypothesis that $b_m = b_1 \neq 0$, $a_\ell = a_0 \neq 0$, and $b_0 = 0$. These matrices are B-normal$(0, 1)$, where in addition, $b_0 = 0$. Therefore, $p_0(\mu) = a_0$ and $q_m(\mu) = b_1 \mu$, and it follows that (20) becomes

$$\bar{\lambda}_i = \frac{a_0}{b_1 \lambda_i}, \qquad \forall \lambda_i \in \Sigma(A),$$

or,

$$\bar{\lambda}_i \lambda_i = \frac{a_0}{b_1}, \qquad \forall \lambda_i \in \Sigma(A).$$

Therefore, $\frac{a_0}{b_1}$ must be positive and real. Matrices that are B-normal$(0, 1)$, with $b_0 = 0$, have eigenvalues that lie on a circle of radius $\sqrt{\frac{a_0}{b_1}}$. That is, they are scaled unitary matrices.

Proof of 6): When A is B-normal(ℓ, m) with $\ell = m$, (19) becomes

$$A^\dagger = \frac{p_m(A)}{q_m(A)} = \frac{a_m}{b_m} + \frac{\tilde{p}_{m-1}(A)}{q_m(A)},$$

where the second equality follows upon division of $p_m(A)$ by $q_m(A)$. Denoting $\hat{A} = (A - \zeta I)$, where $\zeta = \frac{\bar{a}_m}{b_m}$, the above can be rearranged yielding

$$\hat{A}^\dagger = \frac{\tilde{p}_{m-1}(A)}{q_m(A)} = \frac{\hat{p}_{m-1}(\hat{A})}{\hat{q}_m(\hat{A})}.$$

This shows that if A is B-normal(m, m), then there exists a constant ζ, such that $A - \zeta I$ is B-normal$(m - 1, m)$. We apply the results from the $\ell < m$ case to the matrix $\hat{A}$. From this it follows that if A is B-normal(m, m), the only cases that yield more that $m^2 + 1$ distinct eigenvalues are B-normal$(1, 1)$ matrices whose eigenvalues can be obtained by shifting the eigenvalues of a scaled unitary matrix by some constant ζ. These are the shifted unitary matrices described in Section 2. □

The next Theorem gives weaker sufficient conditions on the matrix A for $A \in \mathcal{CG}_B(\ell, m)$.

THEOREM 4.2. $A \in \mathcal{CG}_B(\ell, m)$ if $d(A) \leq \ell + m + 2$, or if there exists polynomials, $p_{\ell'}(\lambda)$ and $q_{m'}(\lambda)$, of degree ℓ' and m' respectively, such that $\ell - \ell' \geq m - m' \geq 0$, and

$$(24) \qquad Q(A) = A^\dagger q_{m'}(A) - p_{\ell'}(A)$$

satisfies

$$\text{Rank}(Q(A)) \leq m - m'.$$

Proof. To show $A \in \mathcal{CG}_B(\ell, m)$, we need to show that there exists coefficients $\{\beta_{i,j}\}_{j=i-m}^{i}$ such that

$$\sigma_{i,k} = \langle \sum_{j=i-m}^{i} \beta_{i,j}\underline{p}_j, A^\dagger \underline{p}_k \rangle_B = 0, \qquad k = 0, ..., i - \ell - 1.$$

There exits polynomials, ψ_k, $\tilde{\psi}_{k-m}$, and $r_{m'-1}^{(k)}$ such that

$$\underline{p}_k = \psi_k(A)\underline{p}_0 = q_{m'}(A)\tilde{\psi}_{k-m'}(A)\underline{p}_0 + r_{m'-1}^{(k)}(A)\underline{p}_0,$$

where the second equality follows from dividing $\psi_k(A)$ by $q_{m'}(A)$. Multiplying through by $A^\dagger$ and then adding and subtracting $p_{\ell'}(A)\tilde{\psi}_{k-m'}(A)\underline{p}_0$ yields

$$A^\dagger \underline{p}_k = (A^\dagger q_{m'}(A) - p_{\ell'}(A))\tilde{\psi}_{k-m'}(A)\underline{p}_0 + p_{\ell'}(A)\tilde{\psi}_{k-m'}(A)\underline{p}_0 + A^\dagger r_{m'-1}^{(k)}(A)\underline{p}_0.$$

Now, since $\ell - \ell' \geq m - m' \geq 0$, we have for $k = 0, ..., i - \ell - 1$ that

$$p_{\ell'}(A)\tilde{\psi}_{k-m'}(A)\underline{p}_0 \in \text{sp}\{\underline{p}_0, \underline{p}_1, ..., \underline{p}_{i-m-1}\},$$

and thus

$$p_{\ell'}(A)\tilde{\psi}_{k-m'}(A)\underline{p}_0 \perp_B \sum_{j=i-m}^{i} \beta_{i,j}\underline{p}_j.$$

Also,

$$A^\dagger r_{m'-1}^{(k)}(A)\underline{p}_0 \in \text{sp}\{A^\dagger \underline{p}_0, A^\dagger \underline{p}_1, ..., A^\dagger \underline{p}_{m'-1}\},$$

and by hypothesis,

$$(A^\dagger q_{m'}(A) - p_{\ell'}(A))\tilde{\psi}_{k-m'}(A)\underline{p}_0 \in \text{Range}(Q(A)).$$

It follows that we only need to choose the $\beta_{i,j}$'s so that

$$(A^\dagger r_{m'-1}(A)\underline{p}_0 + Q(A)\tilde{\psi}_{k-m'}(A)\underline{p}_0) \perp_B \sum_{j=i-m}^{i} \beta_{i,j}\underline{p}_j.$$

Letting $\{\underline{\phi}_j\}_{j=1}^{t}$ be a basis for the Range($Q(A)$), this is equivalent to solving the system

(25)
$$\begin{bmatrix} \langle \underline{p}_i, A^\dagger \underline{p}_0 \rangle_B & \langle \underline{p}_{i-1}, A^\dagger \underline{p}_0 \rangle_B & \cdots & \langle \underline{p}_{i-m}, A^\dagger \underline{p}_0 \rangle_B \\ \vdots & \vdots & & \vdots \\ \langle \underline{p}_i, A^\dagger \underline{p}_{m'-1} \rangle_B & \langle \underline{p}_{i-1}, A^\dagger \underline{p}_{m'-1} \rangle_B & \cdots & \langle \underline{p}_{i-m}, A^\dagger \underline{p}_{m'-1} \rangle_B \\ \langle \underline{p}_i, \underline{\phi}_1 \rangle_B & \langle \underline{p}_{i-1}, \underline{\phi}_1 \rangle_B & \cdots & \langle \underline{p}_{i-m}, \underline{\phi}_1 \rangle_B \\ \vdots & \vdots & & \vdots \\ \langle \underline{p}_i, \underline{\phi}_t \rangle_B & \langle \underline{p}_{i-1}, \underline{\phi}_t \rangle_B & \cdots & \langle \underline{p}_{i-m'}, \underline{\phi}_t \rangle_B \end{bmatrix} \begin{pmatrix} \beta_{i,i} \\ \vdots \\ \beta_{i,i-m'-1} \\ \beta_{i,i-m'} \\ \vdots \\ \beta_{i,i-s} \end{pmatrix} = \begin{pmatrix} 0 \\ \vdots \\ 0 \\ 0 \\ \vdots \\ 0 \end{pmatrix}.$$

The first m' rows correspond to making

$$A^\dagger r_{m'-1}(A)\underline{p}_0 \perp_B \sum_{j=i-m}^{i} \beta_{i,j}\underline{p}_j,$$

while the last t rows correspond to making

$$\text{Range}(Q(A)) \perp_B \sum_{j=i-m}^{i} \beta_{i,j}\underline{p}_j.$$

If $t = \dim(\text{Range}(Q(A))) \leq m - m'$, then the system can always be solved. Thus, $A \in \mathcal{CG}_B(\ell, m)$. □

REMARK 4.2. *As in Lemma 4.1, the system (25) will always have a nontrivial solution, but the algorithm will break down unless there is a solution with $\beta_{i,i} \neq 0$. We believe that, as in the case where A is B-normal(ℓ, m), breakdown occurs for $\underline{p}_0$ in a set of measure zero and can be avoided by a reorganization of the computations. This is a current topic of exploration.*

5 Conclusion

To determine when a conjugate gradient method can be implemented economically, we must consider all possible forms of recursions for the direction vectors. The work by Jagels and Reichel ([5],[6]) demonstrates that the form of recursion given by (1) is not general enough to include all possibilities. In this paper, we have constructed a more general form of recursion called a single (ℓ, m) recursion. Since many multiple recursions can be written equivalently in this form, we began studying multiple recursion conjugate gradient algorithms by analyzing single (ℓ, m) recursions.

Sufficient conditions on the matrix A were determined, in order for a conjugate gradient method to be implemented with a single (ℓ, m) recursion. In the absence of breakdown, if A is B-normal(ℓ, m), or a low rank modification of a B-normal(ℓ, m) matrix, then a conjugate gradient algorithm can be carried out using an (ℓ, m) recursion. For B-normal(ℓ, m) matrices, breakdowns can be avoided by reorganizing the computations. We conjecture that this is also the case for the low rank modifications of B-normal(ℓ, m) matrices in Theorem 4.2. For $\ell, m \leq 2$, the conditions stated in Theorem 4.2 are also necessary. Details will be given in [2].

This work leaves many unanswered questions. For example: necessary conditions for $\ell, m > 2$, and the issue of breakdown in the (ℓ, m) recursion for low rank perturbations of B-normal(ℓ, m) matrices. To consider all possible recursions for the direction vectors, those multiple recursions that cannot be written as single short (ℓ, m) recursions must be characterized. Further research is being done to address some of these issues.

References

[1] S. F. Ashby, T. A. Manteuffel and P. E. Saylor, *A Taxonomy for Conjugate Gradient Methods*, SIAM J. Numer. Anal., 26 (1990), pp. 1542-1568.

[2] T. Barth, Ph.D. thesis, in progress.

[3] V. Faber and T. A. Manteuffel, *Necessary and Sufficient Conditions for the Existence of a Conjugate Gradient Method*, SIAM J. Numer. Analysis, Vol. 21, No. 2, (1984) pp. 352-362.

[4] W. B. Gragg, *Positive definite Toeplitz matrices, the Arnoldi process for isometric operators, and Gaussian quadrature on the unit circle*, J. Comp. Appl. Math, 46 (1993), pp. 183-198.

[5] C. F. Jagels, and L. Reichel, *The isometric Arnoldi process and an application to iterative solution of large linear systems*, in Iterative Methods in Linear Algebra, eds. R. Beauwens and P. de Groen, Elsevier, Amsterdam, 1992, pp. 361-369.

[6] C. F. Jagels, and L. Reichel, *A Fast Minimal Residual Algorithm For Shifted Unitary Matrices*, Numer. Linear Algebra Appl., 1 (1994), pp. 555-570.

[7] W. D. Joubert, and D. M. Young, *Necessary and sufficient conditions for the simplification of generalized conjugate-gradient algorithms*, Linear Algebra Appl., 1988, pp. 449-485.

[8] W. D. Joubert, *Lanczos Methods for the Solution of Nonsymmetric Systems of Linear Equations*, SIAM J. on Matrix Analysis and Appl., vol. 13, no. 3, July 1992, pp. 926-943.

[9] Y. Saad, and M. H. Schultz, *GMRES: a generalized minimal residual algorithm for solving nonsymmetric linear systems*, SIAM J. Sci. Statist. Comput., 1986, No. 7, pp. 856-869.

[10] D. S. Watkins, *Some Perspectives On The Eigenvalue Problem*, SIAM Reviews, Sept. 1993, Vol. 35, No. 3, pp. 430-471.

Chapter 11
Conjugate Gradient Based Implementation of Interior Point Methods for Network Flow Problems*

Sanjay Mehrotra[†]   Jen-Shan Wang[†]

Abstract

This paper suggests enhancements to preconditioned conjugate gradient based implementations of interior point methods for linear network optimization problems. A new preconditioner is proposed and its effectiveness is demonstrated. We propose several additional refinements to the previous implementations, including an adaptive conjugate gradient termination criterion, and a new starting point solution. Our computational results are compared with results from PDNET, DLNET, CS3.3, RELAX-IV and NETOPT (CPLEX3.0).

1 Introduction

The minimum cost network flow (MCNF) problem seeks to minimize the total flow cost in a network while conserving flows (Kirchoff's law) and satisfying the arc capacity constraints. Let $\mathcal{G} = (\mathcal{V}, \mathcal{E})$ be a directed graph with $\mathcal{V} = \{1, \ldots m\}$ representing the vertex set and $\mathcal{E}$ representing the set of n edges. An edge is an ordered pair ij that connects vertex i and vertex j. There is a cost c_{ij} per unit flow on edge ij. The allowable flow on each edge is constrained by the edge capacity, which has a lower bound l_{ij} and an upper bound u_{ij}. Without loss of generality we assume that $l_{ij} = 0 \ \forall ij \in \mathcal{E}$. We also assume that $\mathcal{G}$ is connected. Each vertex has a flow requirement b_i. A vertex is called a source, a sink, or a transshipment vertex if $b_i > 0$, $b_i < 0$, or $b_i = 0$, respectively. The MCNF problem is formulated as follows:

$$\begin{aligned} \text{minimize} \quad & \sum_{ij \in \mathcal{E}} c_{ij} x_{ij} \\ \text{subject to:} \quad & \sum_{ik \in \mathcal{E}} x_{ik} - \sum_{ki \in \mathcal{E}} x_{ki} = b_i, \ i \in \mathcal{V}, \\ & 0 \leq x_{ij} \leq u_{ij}, \ ij \in \mathcal{E}. \end{aligned}$$

Here x_{ij} is the flow on edge $ij \in \mathcal{E}$. It is assumed that the total supply is equal to the total demand, i.e., $\sum_{i \in \mathcal{V}} b_i = 0$, otherwise the problem is infeasible.

The MCNF problem has been studied extensively because of its practical importance and structural properties. The recent book of Ahuja, Magnanti and Orlin [1] provides a comprehensive description of these problems and their applications. Several state of the

---

*Supported in part by the grant CCR-9019469 from the National Science Foundation and the grants N00014-93-1-0317 and N00014-95-1-0743 from the Office of Naval Research.

[†]Department of Industrial Engineering and Management Sciences, Northwestern University, Evanston, IL 60208-3119, U.S.A.

art algorithms for solving MCNF problem are discussed in Chapters 9 to 11 of this book.

Recently there has been considerable interest [18, 19, 9] in studying the prospects of interior point (IP) methods for solving the MCNF problem. The IP methods are known to take very few iterations to solve a linear program. However, each iteration of these methods can be expensive because it must find a search direction by solving the normal equations. The efficient implementations of IP. methods for the MCNF problem as studied in Karmarkar and Ramakrishnan [10], Resende and Veiga [18, 19], Portugal et al. [15], Vaidya and Joshi [9] are based on approximately solving the normal equations using the preconditioned conjugate gradient (PCG) method. An inexpensive and effective preconditioner is required for the PCG based implementations of IP methods to be successful.

Resende and Veiga [18] first reported that a combination of diagonal and maximum weight spanning tree preconditioners (henceforth called RV preconditioner) is effective for assignment problems. They found that the diagonal preconditioner was effective in the earlier phases of the algorithm and the tree preconditioner was effective towards the end. In a subsequent computational work [19] they found that this preconditioner combination is also effective for the general MCNF problems. A heuristic strategy was used to switch from the diagonal to the tree preconditioner.

Portugal et al. [15] suggested a new preconditioner based on an incomplete QR factorization. Portugal et al. [15] showed that their preconditioner was better than RV preconditioner on assignment and transportation problems. However, (in our computational results) it is inferior to the RV preconditioner on general MCNF problems.

This paper makes the following contributions.

- It proposes a new preconditioner. The computational expense of this preconditioner is the same as that of RV preconditioner and the incomplete QR preconditioner. However, it is shown to be effective on all classes of MCNF problems. The improvement varies among various problem classes. On large assignment and transportation problems it requires essentially one PCG iteration per interior point iteration, and for several classes of general MCNF problems it requires on the average two times fewer PCG iterations.

- It proposes a dynamic PCG termination criterion. This criterion appears to stabilize the performance of the dual barrier method (as well as the dual affine scaling method and the primal-dual method) that was implemented for our numerical experiments. The suggested criterion makes significant improvement in the performance of interior point implementations using the proposed new preconditioner as well as the RV preconditioner.

A code, INETS (interior network solver), based on the above improvements is developed. Computational tests with this code are given on **Assignment, Transportation, Grid-8, Grid-Long, Grid-Wide, Mesh-1 and Netgen** problems. Computationally it is found that the above improvements result in an network interior point code which required approximately four times fewer CG iterations and it is estimated to be 2 to 3 times faster than the DLNET and the PDNET implementations of Resende and Veiga described in [16, 18] and Portugal et al. [16].

We have also compared the performance of INETS with other state of the art MCNF solvers: CS3.3 implementing a scaling push-relabel algorithm [4, 6], RELAX-IV implementing the Relaxation algorithm [2, 3], and CPLEX3.0's NETOPT implementing network simplex method. The comparisons are made on a Sun Sparc-10 workstation. Our qualitative conclusions regarding the relative performance of these codes compared with INETS are similar to those of Resende and Veiga [18] and Portugal et al. [16], however the gap between the performance of the specialized network codes and the network interior point implementations is narrowed. More specifically, there was no clear winner between INETS, RELAX-IV and NETOPT. However, CS3.3 was found to be the fastest code on general MCNF problems.

This paper is organized as follows. In the next section an implementation of the dual barrier method is described. For completeness the outline of the dual barrier algorithm is described in Section 2.1, the preconditioned conjugate gradient method is described in Section 2.2 and additional details regarding the starting point and the termination of dual barrier method are given in Section 2.3.

The new preconditioner and its computation is described in Section 3. Sections 4 to 5 are on computational experiments. Section 4 gives the description of the test problems. Section 5.1 compares the performance of the proposed preconditioner with the Resende-Veiga preconditioner and the incomplete QR preconditioner. Section 5.2 compares the performance of INETS with CS3.3, RELAX-IV, and NETOPT.

Notation
$S = diag(s)$ represents a diagonal matrix whose ith diagonal element is s_i. e is taken to be a vector of all ones. x_S represents elements of the vector x that correspond to indices in set S. $\mathcal{T}$ is used to represent the index set of variables corresponding to the edges representing a spanning tree of $\mathcal{G}$. The set $\mathcal{T}$ is also called a spanning tree of $\mathcal{G}$ to simplify discussion. A_S represents columns of A that correspond to column indices in the set S.

2 Interior Point Implementation

Three types of interior point methods have been studied extensively in the recent years. These are the affine scaling methods, the barrier methods [8] and the primal-dual methods [14]. The affine scaling methods and the barrier methods work on either the primal or the dual problem, while the primal-dual methods work on the primal and dual problems together. The global convergence of the affine scaling methods is known for a step size which is a small fraction (≈ 0.67) to the boundary [20]. The barrier methods are related to the affine scaling methods. The barrier methods and the primal-dual methods are globally convergent in polynomial time provided that an appropriate rule is used for updating the barrier parameter [8, 14].

The primal-dual methods have the advantage in the relative ease with which they compute the barrier parameter [12]. The primal-dual methods are also known to have quadratic asymptotic convergence rates [22, 13]. However, the dual methods have the advantage over the primal-dual methods in that algorithmically it is easy to work with an approximate solution of the normal equations solved at each iteration of these methods. Furthermore, it is also possible to work with a subset of edges while solving the normal equations.

The dual barrier method, the dual affine scaling method and the primal-dual method were implemented and extensively tested in our experiments. In these experiments the dual barrier method following the implementation described in this paper was found to be most stable. Typically the number of iterations, the total CG iterations and the CPU time for the dual barrier method as implemented here and the dual affine scaling method were similar. However, while not using the early stopping rules, it was found that the dual affine scaling method (with large step size) failed to converge to an optimal solution on a few occasions. In these experiments we found that the primal-dual methods required a tighter PCG convergence tolerance for the interior point method to converge, which generally resulted in more PCG iterations. A comparison of our results with the truncated Newton primal-dual implementation of Portugal et al. [16] in PDNET indicate that generally the number of interior point iterations and the PCG iterations taken in our implementation of the dual barrier method are better than those taken by PDNET and DLNET. As a result we could not conclude in favor of using a primal-dual algorithm. The dual barrier method as implemented in INETS is now described.

The problem (NF) is rewritten in the following compact form (CF):

$$\text{minimize } \{c^T x | \hat{A}x = \hat{b},\ 0 \leq x \leq u\},$$

where $c, u \in \Re^n$ and $\hat{b} \in \Re^m$. The matrix $\hat{A} \in \Re^{m \times n}$ is called the vertex-edge incidence matrix of $\mathcal{G}$. The matrix $\hat{A}$ has one linearly dependent row. This linear dependence can be removed by deleting any row from $\hat{A}$. Hence the following linear programming problem is considered (LP):

$$\text{minimize } \{c^T x | Ax = b,\ 0 \leq x \leq u\},$$

where $[A:b]$ is obtained from $[\hat{A}:\hat{b}]$ by deleting a row. The dual of (LP) is (LD):

$$\text{maximize } \{b^T y - u^T z | A^T y + s - z = c,\ s, z \geq 0\}.$$

2.1 The Dual Barrier Method.

The dual barrier method considers a sequence of dual barrier problems (DBP):

$$\text{maximize } \{b^T y - u^T z + \mu \sum_{j=1}^{n}[\ln s_j + \ln z_j] \mid A^T y + s - z = c,\ s, z > 0\}$$

for decreasing value of $\mu > 0$. As $\mu \to 0$, the optimal solution of (DBP) approaches the optimal solution of (LP). The dual barrier method computes the Newton direction for (DBP) and takes a step along this direction. A proper control mechanism for the barrier parameter yields a polynomial time algorithm [8].

The Newton Direction.
The first order necessary (and sufficient) conditions for the maximizer of (DBP) are:

(1)
$$\begin{aligned}
A^T y + s - z - c &= 0 \\
Aw - b &= 0 \\
w - \mu S^{-1} e &= 0 \\
-w - \mu Z^{-1} e + u &= 0,
\end{aligned}$$

where $S = diag(s)$, and $Z = diag(z)$, and w is the vector of Lagrange multipliers corresponding to the equality constraints. After eliminating w from (1) we get:

$$
\begin{aligned}
A^T y + s - z - c &= 0 \\
\mu A S^{-1} e - b &= 0 \\
\mu (S^{-1} + Z^{-1}) e - u &= 0.
\end{aligned}
\tag{2}
$$

Let (y^k, s^k, z^k), $s^k > 0, z^k > 0$ be a feasible dual interior solution. The Newton direction (d_y, d_s, d_z) for the problem (DBP) at (y^k, s^k, z^k) is given by

$$
\begin{aligned}
A^T d_y + d_s - d_z &= 0 \\
\mu A (S^k)^{-2} d_s &= \mu A (S^k)^{-1} e - b \\
\mu ((S^k)^{-2} d_s + (Z^k)^{-2} d_z) &= \mu ((S^k)^{-1} + (Z^k)^{-1}) e - u.
\end{aligned}
\tag{3}
$$

Let us define $D^2 = ((S^k)^{-2} + (Z^k)^{-2})^{-1}$. The solution of (3) is given by

$$
\begin{aligned}
d_y &= \tfrac{1}{\mu}(AD^2 A^T)^{-1}(b - A[\mu(S^k)^{-1} e + D^2((Z^k)^2 u - \mu(S^k)^{-1} Z^k (S^k + Z^k) e)]) \\
d_z &= D^2[(Z^k)^2 A^T d_y - \tfrac{1}{\mu}(Z^k)^2 (S^k)^2 D^2 u + Z^k S^k (S^k + Z^k) e] \\
d_s &= d_z - A^T d_y.
\end{aligned}
\tag{4}
$$

The Dual Barrier Iteration.

Let (y^k, s^k, z^k) be the iterate at iteration k. The Newton direction is first computed at (y^k, s^k, z^k) and the next iterate is then generated by

$$
\begin{aligned}
y^{k+1} &= y^k + \alpha d_y \\
s^{k+1} &= s^k + \alpha d_s \\
z^{k+1} &= z^k + \alpha d_z,
\end{aligned}
\tag{5}
$$

where α is a step length in the dual space which ensures $s^{k+1} > 0, z^{k+1} > 0$, and that enough reduction is achieved in the dual barrier function. We take a step of $0.9(\epsilon_{step} = 0.9)$ to the boundary in our implementation. At each step, the barrier parameter μ is reduced. The details of the choice of step length and the reduction in barrier parameter μ differ in various versions of dual barrier methods (for example see Gonzaga [8]). The practical implementations of the method adjust μ heuristically (for example see Vanderbei [21]) and do not perform a line search to ensure reduction in the barrier function value. In INETS the barrier parameter is set as follows. Let

$$
\phi^k = b^T y^k - u^T z^k, \quad \Phi^k = \frac{|\phi^{k-1} - \phi^k|}{\max(1.0, |\phi^k|)} \quad (\Phi^0 = 1)
$$

be the dual objective value and the relative improvement in the dual objective value respectively. At the first iteration $\mu^1 = 0$. After the first iteration

$$
\mu^k = \epsilon_\mu \frac{(\phi^k - \phi^{k-1})}{n}
\tag{6}
$$

where $\epsilon_\mu = 0.01$ is used for computational results in section 5 and 6.

```
0. i = 0
1. ξⁱ = M⁻¹b̄.
2. rⁱ = b̄ − (AD²Aᵀ)ξⁱ.
3. zⁱ = M⁻¹rⁱ.
4. pⁱ = zⁱ.
5. while the stopping criterion is not satisfied do:
6.     qⁱ = (AD²Aᵀ)pⁱ.
7.     αⁱ = (zⁱ)ᵀrⁱ/(pⁱ)ᵀqⁱ.
8.     ξⁱ⁺¹ = ξⁱ + αⁱpⁱ.
9.     rⁱ⁺¹ = rⁱ − αⁱqⁱ.
10.    zⁱ⁺¹ = M⁻¹rⁱ⁺¹.
11.    βⁱ = (zⁱ⁺¹)ᵀrⁱ⁺¹/(zⁱ)ᵀrⁱ.
12.    pⁱ⁺¹ = zⁱ⁺¹ + βⁱpⁱ.
13.    i = i + 1.
```

FIG. 1. *PCG Algorithm*

2.2 Conjugate Gradient (CG) Method

The dual barrier method solves the normal equations

$$(AD^2A^T)d_y = \bar{b}, \tag{7}$$

where $\bar{b} = \frac{1}{\mu}(b - A[\mu(S^k)^{-1}e - D^2((Z^k)^2u + \mu(S^k)^{-1}Z^k(Z^k + S^k)e)])$, to determine the direction d_y. For the network flow problem, the matrix AD^2A^T is sparse, but its Cholesky factor fill-in significantly [19]. It is therefore desirable to consider iterative methods, such as the preconditioned conjugate gradient (PCG) method to solve (7). A matrix M, so that $M^{-1} \approx (AD^2A^T)^{-1}$ and M^{-1} is easily computable, is used as a preconditioner to improve the performance of the CG method. The PCG method is outlined in Figure 1 for completeness. The choice of the preconditioner M and the stopping criterion is discussed in Section 3.

2.3 Starting Solution for the Dual Barrier Method

The initial solution is computed as follows. Let $y^0 = 0$, and for $i = 1, \ldots, n$, let $\tilde{s}_i^0 = c_i, \tilde{z}_i^0 = 0$ if $c_i \geq 0$ and $\tilde{s}_i^0 = 0, \tilde{z}_i^0 = -c_i$ if $c_i < 0$. This gives us a feasible dual solution. A feasible interior solution can be generated by taking $s^0 = \tilde{s}^0 + \beta e$ and $z^0 = \tilde{z}^0 + \beta e$ for any $\beta > 0$.

INETS computes the value of β as follows. It first finds a primal feasible solution by solving a maximum flow problem. INETS calls PRF, an implementation of push-relabel method written by Goldberg and Cherkassky [4] to solve the max-flow problem. If PRF can not find a feasible solution, then the problem is infeasible. Otherwise, let x^0 be the primal feasible solution and take $\beta = \epsilon_{start}(c^Tx^0 - b^Ty^0 + u^T\tilde{z}^0)/2n$. The computational results in this paper use $\epsilon_{start} = 0.5$.

2.4 Termination for the Dual Barrier Method

The dual barrier method needs to generate an optimal primal solution from a dual solution. We follow the approaches described in Resende, Tsuchiya, and Veiga [17] and Resende and Veiga [19], for terminating the algorithm. These are described below.

Spanning Tree Approach. Let $\mathcal{T}$ be a spanning tree of $\mathcal{G}$ and $\mathcal{N}$ be its complement set and $A = [A_\mathcal{T} : A_\mathcal{N}]$. The primal variables corresponding to the edges in $\mathcal{N}$ are set to their bounds as follows:

$$x_i^* = \begin{cases} 0, & \text{if } s_i^k > z_i^k, \\ u_i, & \text{otherwise.} \end{cases}$$

If the solution $x_\mathcal{T}^*$ of $A_\mathcal{T} x_\mathcal{T} = b - A_\mathcal{N} x_\mathcal{N}$ satisfies the bound constraints, x^* is a basic feasible solution. To verify its optimality, a tentative dual solution is constructed by orthogonally projecting the current dual solution y^k onto the dual face $\{y \mid A_\mathcal{F}^T y = c_\mathcal{F}\}$, where $\mathcal{F} = \{i \mid i \in \mathcal{T}, 0 < x_i^* < u_i\}$. Using the fact that $\mathcal{F}$ is a spanning forest, Resende and Veiga [18] show that this projection can be performed efficiently in $O(m)$ time.

Let y^* be the projected dual solution, and $\tilde{c} = c - A^T y^*$. If $\tilde{c}_i > 0$, let $s_i^* = \tilde{c}_i$ and $z_i^* = 0$, else let $s_i^* = 0$ and $z_i^* = -\tilde{c}_i$. If the duality gap $\Psi^* = c^T x^* - b^T y^* + u^T z^*$ equals to zero, then we have an optimal solution pair. Also, if the data are all integers, one can claim optimality if $\Psi^* < 1$.

In our computations the spanning tree used in this approach is the same as the tree used for computing the preconditioner described in Section 3. Since this tree need not represent an optimal feasible basis, this is a heuristic approach. INETS calls the procedure implementing this approach after $\Phi^k \leq 10^{-2}$. The computational experience indicates that this approach is adequate in many cases.

Optimal Partitioning Approach.
The optimal partitioning approach of Resende and Veiga [18] relies on the fact that (LP) and (LD) have one optimal solution pair which is strictly complementary. Several interior point methods converge to one such solution [22]. The dual constraints are partitioned into two sets σ and $\bar{\sigma}$ as follows.

Let $\epsilon_0 < 1, \epsilon_1 < 1$ and $\sigma = \bar{\sigma} = \emptyset$.
For $i = 1, \ldots, n$, do
{
    If $z_i^k/z_i^{k-1} > \epsilon_0$ and $s_i^k/s_i^{k-1} < \epsilon_1$,
      set x_i^* to its lower bound and $\bar{\sigma} = \bar{\sigma} \cup \{i\}$.
    If $z_i^k/z_i^{k-1} < \epsilon_1$ and $s_i^k/s_i^{k-1} > \epsilon_0$,
      set x_i^* to its upper bound and $\bar{\sigma} = \bar{\sigma} \cup \{i\}$.
    Otherwise, $\sigma = \sigma \cup \{i\}$.
}

A maximum weighted spanning forest $\mathcal{F}$ is formed from the edges in σ and a tentative dual solution y^* is constructed by orthogonally projecting the current dual solution y^k onto the dual face $\{y \mid A_{\mathcal{F}}^T y = c_{\mathcal{F}}\}$. The solution y^* is used to compute s^* and z^* as described in the spanning tree approach.

A primal solution is generated as follows. Let $\sigma^* = \{i \mid |c_i - A_i^T y^*| < \epsilon_{partition}, i = 1, \ldots, n\}$, and $\bar{\sigma}^* = \{1, \ldots, n\} \backslash n$. Here ϵ is a small tolerance (INETS uses $\epsilon_{partition} = 10^{-8}$ as default). Now for $i \in \bar{\sigma}^*$ set

$$x_i^* = \begin{cases} 0, & \text{if } c_i - A_i^T y^* > 0, \\ u_i, & \text{otherwise.} \end{cases}$$

Next, we define a feasibility problem:

$$A_{\sigma^*} x_{\sigma^*} = b - A_{\bar{\sigma}^*} x_{\bar{\sigma}^*},$$

$$0 \leq x_i \leq u_i, i \in \sigma^*.$$

If a feasible solution to the above problem can be found, then the complementary slackness conditions ensure the optimality of primal and dual solutions. This feasibility problem can be solved by solving a maximum flow problem. INETS calls PRF [4] to perform these computations.

Since the cost of computing the maximum flow solution is significant, the procedure implementing the optimal partitioning approach is not performed at every iteration. INETS calls the procedure implementing this approach if the primal objective value corresponding to the spanning tree generated while computing our preconditioner is same for the two successive iterations, or at an iteration where Φ^k is reduced by a factor of 10^{-1} after it has become smaller than 10^{-6}. $\beta_0 = 0.7$ and $\beta_1 = 0.9$ are used in INETS.

3 Implementation of PCG method

Resende-Veiga Preconditioner

The basis in the network simplex method correspond to a spanning tree of the graph $\mathcal{G}$. A nondegenerate MCNF problem has a unique optimal solution and a corresponding optimal spanning tree represented by $\mathcal{T}^*$. Let $A = [A_{\mathcal{T}^*} : A_{\mathcal{N}^*}]$, where $\mathcal{N}^* = \mathcal{G} \backslash \mathcal{T}^*$. Let $(y^k, s^k, z^k) \to (y^*, s^*, z^*)$ in the dual barrier method. Then it is easy to see that $De \to (D_{\mathcal{T}^*} e, 0)$ and the matrix $AD^2 A^T \to A_{\mathcal{T}^*} D_{\mathcal{T}^*}^2 A_{\mathcal{T}^*}^T$. Hence one may expect that $A_{\mathcal{T}} D_{\mathcal{T}}^2 A_{\mathcal{T}}^T$, where $\mathcal{T}$ is a maximum weighted spanning tree, would serve as a good preconditioner as the algorithm converge. This provides a motivation for using $A_{\mathcal{T}} D_{\mathcal{T}}^2 A_{\mathcal{T}}^T$ as a preconditioner. We call this the spanning tree preconditioner.

Most MCNF problems are not non-degenerate. Nevertheless, Resende and Veiga [18, 19] found that the spanning tree preconditioner is very effective in the later iterations of the dual affine scaling algorithm. In the earlier iterations they used the diagonal preconditioner, and switched to the tree preconditioner when the diagonal preconditioner became ineffective. This strategy is called the RV preconditioner.

Several algorithms including Kruskal's algorithm, Prim's algorithm and Sollin's algorithm for computing a maximum weight spanning tree are described in Ahuja, Magnanti and Orlin [1] . Resende and Veiga [18] compute an inexact spanning tree by approximately sorting the edge weights using bucket sort followed by an implementation of Kruskal's algorithm. We follow a similar strategy and typically accomplished this task in $O(n)$ steps for each call of the PCG method. Since solves involving spanning tree can be performed in $O(m)$ time, steps 1, 3 and 10 of PCG method in Figure 2.2 can be performed in $O(m)$ time. Steps 2,4,7,8,9,11,12 are also be performed in $O(m)$ time. Step 6 is performed in $O(n)$ time. Therefore, one PCG iteration is performed in $O(n)$ time.

Incomplete QR Preconditioner

Portugal et al. [15] observed that when the interior point algorithm makes a transition between using the diagonal preconditioner and the tree preconditioner, it is possible that both of these preconditioners do not work well. To circumvent this situation they propose a new preconditioner based on computing an incomplete QR factorization of the matrix DA. They showed that their preconditioner out performed the RV preconditioner for assignment and transportation problems. The computation of Incomplete QR preconditioner suggested by Portugal et al. [15] is described next with the help of an example.

Let $\mathcal{T}$ be a maximum weighted spanning tree of $\mathcal{G}$ with weights D. Let $A = [A_\mathcal{T} : A_\mathcal{N}]$, where $\mathcal{N} = \mathcal{G}\backslash\mathcal{T}$. Assume that we are given a network with 4 nodes and 7 arcs and the matrix $\bar{A}^T = DA^T$ is given below.

$$\bar{A}^T = \begin{bmatrix} D_\mathcal{T} A_\mathcal{T}^T \\ D_\mathcal{N} A_\mathcal{N}^T \end{bmatrix} = \begin{bmatrix} d_5 & & & -d_5 \\ & d_2 & -d_2 & \\ & & -d_4 & d_4 \\ & & & d_7 \\ \cdots & \cdots & \cdots & \cdots \\ -d_1 & & d_1 & \\ d_3 & & & -d_3 \\ & d_6 & -d_6 & \end{bmatrix}$$

The element (5,1) in $\bar{A}^T$ is eliminated first while allowing no fill-in in position (1,3) and (5,4). As a consequence we have

$$\begin{bmatrix} \sqrt{d_5^2 + d_1^2} & & & -\frac{d_5^2}{\sqrt{d_5^2+d_1^2}} \\ & d_2 & -d_2 & \\ & & -\sqrt{d_4^2 + d_1^2} & d_4 \\ & & & d_7 \\ \cdots & \cdots & \cdots & \cdots \\ 0 & & 0 & \\ d_3 & & & -d_3 \\ & d_6 & -d_6 & \end{bmatrix}$$

The same process is repeated for the elements (6,1) and (6,2) which results in the matrix:

$$\begin{bmatrix} \sqrt{d_5^2+d_1^2+d_3^2} & & & -\frac{d_5^2}{\sqrt{d_5^2+d_1^2+d_3^2}} \\ & \sqrt{d_2^2+d_6^2} & -\frac{d_2^2}{\sqrt{d_2^2+d_6^2}} & \\ & & -\sqrt{d_4^2+d_1^2+d_6^2} & \frac{d_4^2}{\sqrt{d_4^2+d_1^2+d_6^2}} \\ & & & \sqrt{d_7^2+d_3^2} \\ \cdots & \cdots & \cdots & \cdots \\ 0 & & 0 & \\ 0 & & & 0 \\ & 0 & 0 & \end{bmatrix}.$$

Because of the form of matrix $D_T A_T^T$, the incomplete QR preconditioner can also be computed in $O(n)$ time after the maximum weight spanning tree is identified.

New Preconditioner
The new preconditioner, which is based on computing an incomplete Cholesky factor of AD^2A^T, combines the ideas of Resende and Veiga [18] and Portugal et al. [15]. Let $\mathcal{T}$ be a maximum weighted spanning tree of $\mathcal{G}$ with edge weights De. Let $A = [A_\mathcal{T} : A_\mathcal{N}]$, $\mathcal{N} = \mathcal{G} \backslash \mathcal{T}$. Let

(8) $$AD^2A^T = \left[A_\mathcal{T} D_\mathcal{T}^2 A_\mathcal{T}^T + A_\mathcal{N} D_\mathcal{N}^2 A_\mathcal{N}^T \right] \approx A_\mathcal{T} D_\mathcal{T}^2 A_\mathcal{T}^T + \Lambda,$$

where Λ is a nonnegative diagonal matrix. The next proposition shows that the Cholesky factors of the matrix $A_\mathcal{T} D_\mathcal{T}^2 A_\mathcal{T}^T + \Lambda$ is easily computable.

PROPOSITION 3.1. *The Cholesky factor L, such that $LL^T = A_\mathcal{T} D_\mathcal{T}^2 A_\mathcal{T}^T + \Lambda$ can be computed in $O(m)$ computational steps.*

Proof. We observe that at each stage of factorization of $A_\mathcal{T} D_\mathcal{T}^2 A_\mathcal{T}^T + \Lambda$, we have a column with exactly one off-diagonal nonzero element. Therefore, computation of the first column of L is done in $O(1)$ time. The computation for the second and subsequent columns proceed in the same way. □

There is a flexibility in the choice of Λ and we experimented with different possible ways of computing it. A simple choice which has worked well in practice is to take

$$\Lambda = \rho \times tr(A_\mathcal{N} D_\mathcal{N}^2 A_\mathcal{N}^T),$$

where $tr(A_\mathcal{N} D_\mathcal{N}^2 A_\mathcal{N}^T)$ represents the diagonal elements in the matrix $A_\mathcal{N} D_\mathcal{N}^2 A_\mathcal{N}^T$ and ρ is computed adaptively. The new preconditioner can be viewed as adaptively switching between the diagonal preconditioner and the tree preconditioner in the RV preconditioner, while using the tree nonzero structure information as in the case of the incomplete QR preconditioner computation of Portugal et al. [15].

The computational results described in the next section take $\rho = 1$ for `assignment`, `transportation` and `netgen` problems. It uses $\rho = 0.1 * \min\{D_\mathcal{T}\}/\max\{D_\mathcal{T}\}$ for the `grid` problems, `grid-long` problems, and `grid-wide` problems, and $\rho = 10 * \min\{D_\mathcal{T}\}/\max\{D_\mathcal{T}\}$ for the `mesh` problems.

CG Termination Criterion

Let $\tilde{d}_y$ be an approximate solution of (7) and

(9) $$\cos\theta = \frac{|\bar{b}^T(AD^2A^T)\tilde{d}_y|}{\|\bar{b}\|\cdot\|(AD^2A^T)\tilde{d}_y\|}.$$

Following Karmarkar and Ramakrishnan [10] and Resende and Veiga [18], we terminate PCG iterations when $|1 - \cos\theta| \leq \epsilon$. However, instead of using a fixed tolerance of $\epsilon = 10^{-3}$ (as used in Resende and Veiga [18]), ϵ is computed adaptively as follows:

$$\epsilon^k = \begin{cases} \epsilon_{cg}, & \text{if } \Phi^k > 0.1, \\ \epsilon_{cg}\Phi^k, & \text{otherwise.} \end{cases}$$

The computational results reported in the next section use $\epsilon_{cg} = 10^{-3}$. Although a smaller choice (say $5.0 * 10^{-3}$) improves the performance on the test problems, we present results for this tighter choice so that a comparison can be made with the results reported by Resende and Veiga [18].

4 Test Problems

The test problems were generated by using various generators available from DIMACS [5]. Problems of similar dimensions were used for the computational results reported by Resende and Veiga [18], and Portugal et al. [16]. We describe these problems below. We have chosen the size of our test problems so that comparisons could be made with DLNET and PDNET.

Assignment Problems. The `assignment` problems are generated by using generator `assign.c` written by McGeoch [5]. The numbers of supply nodes are set to be equal to the number of demand nodes and there is no transshipment node. If a node is a supply node, $b_i = 1$; if it is a demand node, $b_i = -1$. In our test problems, each supply node is set to connect to 64 demand nodes and the maximum cost used for all the test problems is 1024. The random seed 270001 is used.

Transportation Problems. In this problem class, problems are generated by using a C version of the **NETGEN** network generator developed by Klingman, Napier, and Stutz [11]. In the transportation network, there are no transshipment nodes. The 15 parameters used in **NETGEN** are set as follows:

seed	Random number seed	270001
problem	Problem number (for output)	1
nodes	Number of vertices	2^N
sources	Number of source nodes	2^{N-1}
sinks	Number of sink nodes	2^{N-1}
density	Number of edges (directed)	2^{N+4}
mincost	Minimum arc cost	0
maxcost	Maximum arc cost	4096
supply	Total supply	$2^{(N+2)}$
tsources	Transshipment sources	0
tsinks	Transshipment sinks	0
hicost	Percentage of skeleton arcs with max cost	100%
capacitated	Percentage of arcs set to be capacited	100%
mincap	Minimum arc capacity	1
maxcap	Maximum arc capacity	16384

Problems with $N = 11$ to 14 are generated for use as test problems.

Grid Problems. This class of problems are generated by using Goldberg's goto.c network generator [5]. We generated examples for $n = 8m$ (grid-8 problems), i.e., the total number of arcs equals to 8 times the the number of nodes. In all test problems, arc costs are between 0 and 4096 and the arc capacities are between 0 and 16384. The seed for random number generator is 270001. We also generated grid-16 problems ($n = 16m$), and grid-increasing problems ($n = m^{1.5}$). Computational results with these problems are not included here as they are similar to those for grid-8 problem.

Grid-Long and Grid-Wide Problems. The two problem sets are generated by using Resende's ggraph.f generator [5]. In our grid-long test problems, the grid structures have fixed height 16 and with width from 32 to 4096. grid-wide problems are set to have height 32 to 4096 and a fixed width 16. Maximum cost and maximum edge capacity are set to 10000 for both grid-long problems and grid-wide problems. The random seed is 270001.

Mesh Problems. These problems are generated by using generator mesh1.c, which is a minor modification of Goldberg's mesh.c generator. The graphs are based on grids embedded on a torus. In our test problems, each vertex connects to 2 nearest neighbors (mesh-1). Costs are generated uniformly between -1000 and 1000. Capacities are generated between -1000 and 1000. The results for problem class mesh-4 and mesh-8 are similar and are not included here.

Netgen-Hi and Netgen-Lo Problems. This class of problems are generated by using generator NETGEN [11]. The difference between netgen-hi and netgen-lo is the input parameter maxcap which defines the maximum edge capacity. For netgen-hi problems, parameter maxcap is set to 16384 while it is set to 16 for netgen-lo problems. Except for the following changes, the parameter setting for the transportation problems was used.

sources	Number of source nodes	2^{N-2}
sinks	Number of sink nodes	2^{N-2}
density	Number of edges (directed)	2^{N+3}
supply	Total supply	$2^{2(N-2)}$

Problems with $N = 12$ to 15 are generated.

5 Computational Results

We now describe the computational environment in which computational comparisons were made with other state of the art network solvers. INETS, CS3.3, and NETOPT are written in C. RELAX-IV is written in FORTRAN-77. We ran these codes on a Sun Sparc-10 workstation. The CPU time reported in the subsequent sections are the times required to solve a problem. They do not include the time to read the data. The times for CS3.3, RELAX-IV and NETOPT are the times reported by these codes. The problems for which we report our results are the largest problems we could put on our machine before running out of memory.

5.1 Comparing Different Preconditioners

In this section we compare the performance of the new preconditioner with the RV preconditioner and the incomplete QR preconditioner. Since PDNET and DLNET are not available in public domain, we have taken the performance statistics reported in Resende and Veiga [18] and Portugal et al. [16] for comparison. We also ran our implementation of RV preconditioner to make a direct comparison between the performance of the two preconditioners. These results are reported in Table (1). This table gives information on problem types and problem sizes, interior point iterations, total PCG iterations, average PCG iterations and the CPU times for INETS with new and RV preconditioner, and statistic from [18, 16]. The last two columns of the table list the ratio of total PCG iteration between DLNET and INETS (with new preconditioner), and for the NEW and RV preconditioner within INETS.

The results for assignment and transportation problem are not reported in [18, 16] for DLNET and PDNET, hence are left as blanks in Table (1). From our implementation of RV preconditioner, however, one can see that the total PCG iterations and the CPU times for the new preconditioner is significantly better for these problems. Figures (2–3) provide more detailed comparison on the performance of various preconditioners. These figures show the PCG iterations for our new preconditioner, the RV preconditioner, and the incomplete QR preconditioner for each iteration of the interior point method.

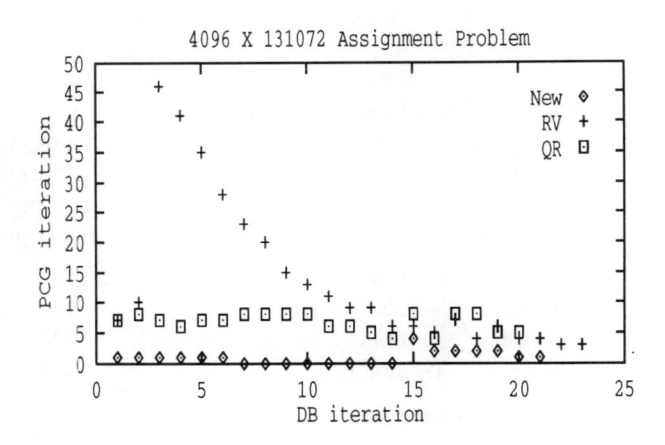

FIG. 2. *Assignment Problem Example*

From Table (1) we conclude that INETS needs fewer PCG iterations than both DLNET and PDNET except for the problem class `grid-long`. The conjugate gradient iterations for `Problem 19` from this problem set are shown in (4), which is typical of the performance of other problems in this class. We find that the performance of the proposed preconditioner was worse in the first 10 iterations. The RV preconditioner for these problems switches to the tree preconditioner at the very first interior point iteration. For these problems a significantly better performance is observed if we increase our initial PCG termination tolerance from 10^{-3} to $5*10^{-3}$. The performance of incomplete QR preconditioner is worse than that of RV and the New preconditioner for all classes of general network problems.

TABLE 1
Performance results for the new preconditioner

	prob size			INETS (NEW)			INETS (RV)			DLNET		PDNET			
	$|\mathcal{V}|$	$|\mathcal{E}|$	iter	cg(av)	time	iter	cg	time	iter	cg	iter	cg	D/N	R/N
						assignment problems								
1	2048	65536	17	7(0.4)	19.2	18	116	27.6						16.6
2	4096	131072	21	20(1.0)	49.3	23	315	107.0						15.8
3	8192	262144	24	17(0.7)	113.8	24	309	220.2						18.2
4	16384	524288	26	31(1.2)	286.2	28	470	745.3						15.2
						transportation problems								
5	256	4096	21	32(1.5)	1.4	24	87	1.6						2.7
6	1024	16430	30	28(0.9)	8.6	31	144	10.0						5.1
7	4096	65766	49	35(0.7)	68.2	50	473	110.2						13.5
8	16384	263266	59	104(1.8)	411.1	57	1372	1055.1						13.2
						grid-8 problems								
9	256	2048	24	123(5.1)	1.0	24	125	1.0	21	181	30	146	1.5	1.0
10	1024	8192	31	136(4.4)	6.0	31	138	5.9	29	234	43	167	1.7	1.0
11	4096	32768	34	247(7.3)	34.2	34	217	31.5	32	376	44	297	1.5	0.9
12	16384	131072	40	275(6.8)	201.6	40	263	217.7	45	615	43	463	2.2	1.0
						mesh-1 problems								
13	1024	2048	18	112(1.9)	1.5	18	190	2.1	36	454	23	200	4.1	1.7
14	4096	8192	24	150(6.3)	10.1	23	358	15.7	64	817	33	425	5.4	2.4
15	16384	32768	27	246(9.1)	77.9	27	902	195.1	52	905	55	1170	3.7	3.7
16	65536	131072	32	360(11.3)	503.5	31	1902	1978.5	83	1183	65	2338	3.3	5.3
						grid-long problems								
17	1026	2000	33	399(12.1)	4.1	30	360	4.0	29	402	31	385	1.0	0.9
18	4098	7952	48	1220(25.4)	44.1	48	1032	41.1	73	1112	46	992	0.9	0.8
19	16386	31760	54	3454(63.9)	565.3	55	2050	378.3	84	2517	136	2301	0.7	0.6
20	65538	126992	179	7534(42.1)	7489.4	176	7032	7639.5	169	4228	103	4931	0.6	0.9
						grid-wide problems								
21	1026	2096	27	177(6.6)	2.3	27	207	2.7	30	391	31	223	2.2	1.2
22	4098	8432	33	193(5.8)	13.3	33	261	14.9	38	536	50	630	2.8	1.4
23	16386	33776	53	247(4.7)	90.3	49	347	108.4	49	458	76	812	1.9	1.4
24	65538	135152	88	375(4.3)	752.4	77	555	827.2	71	388	85	695	1.0	1.5
						netgen-lo problems								
25	256	2048	16	59(3.7)	0.7	16	200	1.1	19	233	21	244	3.9	3.4
26	1024	8214	24	93(3.9)	4.6	24	252	6.5	36	532	33	399	5.7	2.7
27	4096	32858	35	319(9.1)	41.8	34	402	45.3	42	644	49	743	2.0	1.3
28	16384	131409	49	1994(40.7)	734.7	46	823	434.1	73	1220	49	1255	0.6	0.4
						netgen-hi problems								
29	256	2048	20	39(2.0)	0.7	22	107	0.9	29	255	69	281	6.5	2.7
30	1024	8214	33	56(1.7)	5.5	34	251	7.6	41	537	97	377	9.6	4.5
31	4096	32858	42	57(1.4)	33.7	41	402	51.7	61	989	88	489	17.4	7.1
32	16384	131409	44	83(1.9)	162.1	45	788	380.6	74	2012	67	1096	24.2	9.5

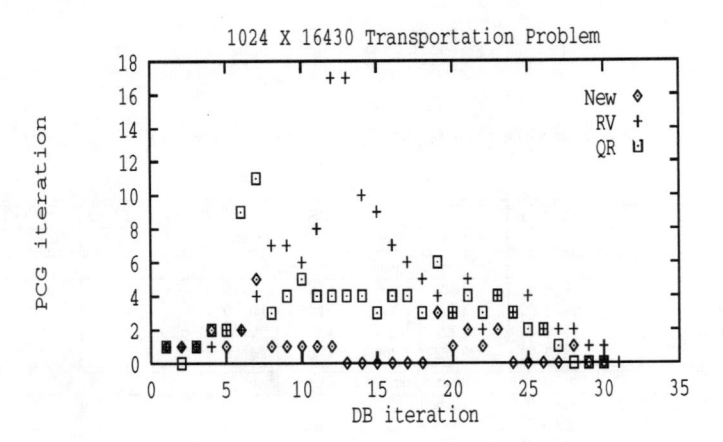

FIG. 3. *Transportation Problem Example*

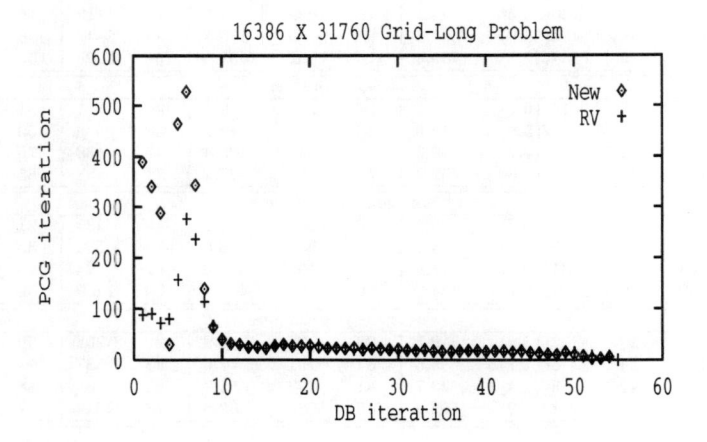

FIG. 4. *Grid-long Problem Example*

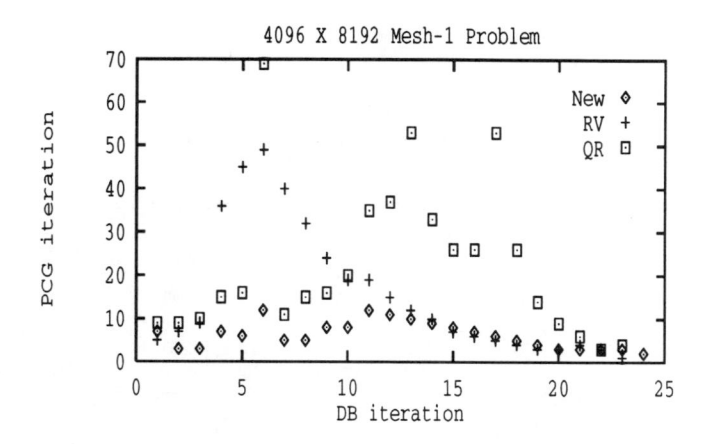

FIG. 5. *Mesh Problem Example*

For most of the problems we tested, INETS also needs fewer interior point iterations than DLNET and PDNET. As an extreme example, INETS consistently needs less than half of the interior point iterations for `mesh-1` problems than DLNET. In Figure (5) we give the PCG iterations for the `mesh` problem to show the improvements due to the new preconditioner.

5.2 Comparing INETS with other solvers

In this section, we compare INETS with CS3.3, RELAX-IV, and NETOPT. NETOPT implements the network simplex method. CS3.3 implements a scaling push-relabel algorithm [6, 7], and RELAX-IV implements the Relaxation algorithm [2, 3]. For assignment problems the results are given using the CSAS code, which implements a variant of push-relabel algorithm specially designed for these problems problem [7]. We point out that INETS, CS3.3, CSAS and NETOPT are written in C programming language, and RELAX-IV is written in FORTRAN. The choice of programming language and a different platform for computing may change some performance ratios. We have used the default setting for CS3.3 and NETOPT. RELAX-IV was run with the two provided options. The smaller of the CPU times from these two options is reported.

Table (2) below shows the results from different solvers. The last three columns of this table gives the ratio of CPU times for INETS and CS3.3 (INETS/CSAS for assignment problems), INETS and RELAX-IV and INETS and NETOPT. While computing the performance ratios discussed below we have ignored the smallest problem from each problem class. RELAX-IV did not return with a solution for the 65538 node `grid-wide` problem. For the 65538 node `grid-long` problem it prints the number of iterations taken by the algorithm as 116, which appears to be incorrect.

From this table, it is clear that CS3.3 is currently the fastest code for solving most of the problem classes. On average it is about 10 times faster than INETS. However, for all

TABLE 2
Comparison with other network solvers

	prob size		INETS		CSAS	RELAX-IV		NETOPT				
	$|\mathcal{V}|$	$|\mathcal{E}|$	iter	time	time	iter	time	iter	time	I/C	I/R	I/N
					assignment problem							
1	2048	65536	17	19.2	1.2	1714	1.8	51735	16.9	16.0	10.7	1.1
2	4096	131072	21	49.3	2.6	4228	6.9	83965	49.37	19.0	7.1	1.0
3	8192	262144	24	113.8	5.3	8474	13.4	281780	196.5	21.5	8.5	0.6
4	16384	524288	26	286.2	12.2	17766	32.1	918169	752.0	23.5	8.9	0.4
	prob size		INETS		CS	RELAX-IV		NETOPT				
	$|\mathcal{V}|$	$|\mathcal{E}|$	iter	time	time	iter	time	iter	time	I/C	I/R	I/N
					transportation problem							
5	256	4096	21	1.4	0.2	661	0.1	1576	0.3	7.0	14	4.7
6	1024	16430	30	8.6	1.7	3569	0.7	7060	2.6	5.1	12.3	3.3
7	4096	65766	49	68.2	14.2	14766	10.1	50211	46.2	4.8	6.8	1.5
8	16384	263266	59	411.1	87.5	57952	91.0	217039	1440.7	4.7	4.5	0.3
					grid-8 problem							
9	256	2048	24	1.0	0.3	2022	0.6	4331	0.4	3.3	1.7	2.5
10	1024	8192	31	6.0	2.2	5549	7.2	33490	9.9	2.7	0.8	0.6
11	4096	32768	34	34.2	20.2	39491	186.8	183357	198.7	1.7	0.2	0.2
12	16384	131072	40	201.6	158.8	245756	3713.6	852100	6656.2	1.3	0.05	0.03
					mesh-1 problem							
13	1024	2048	18	1.6	0.4	5641	0.3	830	0.1	4.0	5.3	16
14	4096	8192	24	10.1	3.2	23414	2.9	16213	9.1	3.2	3.5	1.1
15	16384	32768	27	77.9	26.3	86505	30.2	67151	170.2	3.0	2.6	0.5
16	65536	131072	32	503.5	148.0	382002	226.7	287705	3143.3	3.4	2.2	0.2
					grid-long problem							
17	1026	2000	33	4.1	0.3	2519	0.5	2302	0.3	13.7	8.2	13.7
18	4098	7952	48	44.1	1.7	11873	8.4	8120	5.9	25.9	5.3	7.5
19	16386	31760	54	565.3	12.4	52053	144.6	30390	132.3	45.6	3.9	4.3
20	65538	126992	179	7489.4	80.3	116?	4516.1	90876	2167.6	93.3	1.7	3.5
					grid-wide problem							
21	1026	2096	27	2.5	0.4	5059	0.5	2517	0.2	6.3	5.0	12.5
22	4098	8432	33	13.3	2.8	12939	11.1	10614	1.4	4.8	1.2	9.5
23	16386	33776	53	90.3	23.4	54246	108.6	37069	11.5	3.9	0.8	7.9
24	65538	135152	88	752.4	150.2			143404	74.2	5.0		10.1
					netgen-lo problem							
25	256	2048	16	0.7	0.1	1256	0.1	2671	0.2	7.0	7.0	3.5
26	1024	8214	24	4.8	1.0	5197	1.0	16629	2.7	4.8	4.8	1.8
27	4096	32858	35	47.6	12.8	18680	11.8	68405	35.3	3.7	4.0	1.3
28	16386	131409	49	812.6	106.4	100572	215.7	259576	628.5	7.6	3.8	1.3
					netgen-hi problem							
29	256	2048	20	0.7	0.1	824	0.1	1167	0.2	7.0	7.0	3.5
30	1024	8214	33	5.5	0.9	3517	0.5	7044	1.2	6.1	11.0	4.6
31	4096	32858	42	33.7	7.6	12866	6.6	30798	18.9	4.4	5.1	1.8
32	16384	131409	44	162.1	48.8	60428	46.9	143458	569.0	3.3	3.5	0.3

the problems except the **assignment** and the **grid-long** problems the performance ratio of the two methods show a favorable trend towards INETS with increasing problem size. For assignment problems there is no clear trend, however, for the **grid-long** problems the trend is in favor of CS3.3.

The performance of INETS compared with RELAX-IV is mixed. INETS is about 8 times slower for the **assignment** problems and **transportation** problems, 3 times slower for the **mesh-1** problems, 5 times slower for the **grid-long** problems, and about 5 times slower for the netgen problems. It is about 7 times faster on **grid-8** and competitive on **grid-wide** problems. For all the problems the trend is in favor of INETS with increasing problem size.

Qualitatively the performance comparisons of INETS with NETOPT are similar to those of INETS with RELAX-IV. NETOPT is faster than INETS on **grid-long**, **grid-wide**, and **netgen-lo** problems; and slower on **grid-8** problems. The performance on **assignment**, **transportation**, **mesh-1**, and **netgen-hi** problems is mixed, with the trend in favor of INETS.

References

[1] R. K. Ahuja, T. L. Magnanti, and J. B. Orlin, *Network Flows: Theory, Algorithms, and Applications*, Prentice Hall, Englewood Cliffs, NJ, (1993).

[2] D. P. Bertsekas and P. Tseng, *The relax codes for linear minimum cost network flow problems*, in FORTRAN Codes for Network Optimization, B. Simeone, et al. Eds. , Annals of Operations Research, vol. 13 (1988), pp. 125–190.

[3] ———, *Relax - IV: A Faster Version of the RELAX Code for Solving Minimum Cost Flow Problems*, Tech. Report, Department of Computer Science, M. I. T. , November, 1994.

[4] B. V. Cherkassky and A. V. Goldberg, *On Implementing Push-Relabel Method for The Maximum Flow Problem*, Tech. Report, Computer Science Department, Stanford University, Stanford, CA 94305, USA, September, 1994.

[5] DIMACS, *The First DIMACS international algorithm implementation challenge: the benchmark experiment*, DIMACS, Rutgers University, New Brunswick, NJ, USA, 1991.

[6] A. V. Goldberg, *An Efficient Implementation Of A Scaling Minimum-Cost Flow Algorithm*, Tech Report, Computer Science Department, Stanford University, Stanford, CA 94305, USA, August, 1992.

[7] A. V. Goldberg and R. Kennedy, *An Efficient Cost Scaling Algorithm For The Assignment Problem*, Tech. Report, Computer Science Department, Stanford University, Stanford, CA 94305, USA, July, 1993.

[8] C. C. Gonzaga, *Path following methods for linear programming*, SIAM Review, vol. 34, no. 2 (1992), pp. 167–227.

[9] A. Joshi, A. S. Goldstein, and P. M. Vaidya, *A Fast Implementation of A Path-Following Algorithm for Maximizing a Linear Function Over A Network Polytope*, in Network Flow and Matching: First DIMACS Implementing Challenge, D. S. Johnson and C. C. McGeoch Eds. , DIMACS Series in Discrete Mathematics and Theoretical Computer Science, vol. 12, American Mathematical Society, 1991.

[10] N. K. Karmarkar and K. G. Ramakrishnan, *Computational Results of an Interior Point Algorithm for Large Scale Linear Programming*, Mathematical Programming, vol. 52 (1991), pp. 555–586.

[11] D. Klingman, A. Napier, and J. Stutz, *A Program for generating large scale capacitated assignment, transportation, and minimum-cost flow network problems*, Management Science, vol. 20 (1974), pp. 814–820.

[12] S. Mehrotra, *On the Implementation of the a Primal-Dual Interior Point Method*, SIAM J. Optimization, vol. 2, no. 4 (1992), pp. 575–601.
[13] ———, *Quadratic Convergence in a Primal-Dual Method*, Mathematics of Operations Research, vol. 18 (1993), pp. 741–751.
[14] S. Mizuno, M. J. Todd, and Y. Ye, *On adaptive-Step Primal-Dual Interior Point Algorithms for Linear Programming*, Mathematics of Operations Research, vol. 18 (1993), pp. 964–981.
[15] L. Portugal, F. Bastos, J. Judice, J. Paixao, and T. Terlaky, *An Investigation of Interior Point Algorithms For The Linear Transportation Problem*, Tech. Report, Department of Mathematics, University of Coimbra, Coimbra, Portugal, 1993.
[16] L. Portugal, M. G. C. Resende, G. Veiga, and J. Judice, *A Truncated Primal-Infeasible Dual-Feasible Network Interior Point Method*, Tech. Report, AT&T Bell Laboratories, Murray Hill, NJ, USA, 1994.
[17] M. G. C. Resende, T. Tsuchiya, and G. Veiga, *Identifying The Optimal Face Of A Network Linear Program With A Globally Convergent Interior Point Method*, Large Scale Optimization: State of the Art, D. W. Hearn and P. M. Pardalos, Eds. , Kluwer (1994), pp. 362–387.
[18] ———, *An Efficient Implementation of a Network Interior Point Method*, Tech. Report, AT&T Bell Laboratories, Murray Hill, NJ, USA, 1992.
[19] ———, *An Implementation of the Dual Affine Scaling Algorithm for Minimum Cost Flow on Bipartite Uncapacitated Networks*, SIAM Journal on Optimization, vol. 3, (1993), pp. 516–537.
[20] T. Tsuchiya and M. Muramatsu, *Global Convergnce of The Long-Step Affine Scaling Algorithm For Degenerate Linear Programming Problems*, Tech. Report, The Institute of Statistical Mathematics, Tokyo, Japan, January, 1992.
[21] R. J. Vanderbei, *ALPO: Another Linear Program Optimizer*, ORSA Journal on Computing, vol. 5, no. 2 (1993), pp. 134–146.
[22] Y. Ye, O. Güler, R. A. Tapia, and Z. Zhang, *A Quadratically Convergent $O(\sqrt{n}L)$-iteration algorithm for Linear Programming*, Mathematical Programming, vol. 59, no. 2 (1993), pp. 151–162.

Chapter 12
SOR as a Parallel Preconditioner[*]

M. A. DeLong[†]   J. M. Ortega[†]

Abstract

In this paper we give the results of numerical experiments using k steps of the SOR iteration as a preconditioner for GMRES for solving nonsymmetric linear systems of equations. The two machines considered are the Intel Paragon and IBM SP-2, using 16 processors on both. We show that the SOR-GMRES method has good overall parallel efficiency, although the Gram-Schmidt portion of the GMRES algorithm degrades the excellent parallel efficiency of the preconditioning. The parallel efficiencies on the SP-2, although still good, are significantly below those on the Paragon.

1 Introduction

The conjugate gradient (CG) method (see, e.g. Golub and Van Loan[5]) for solving a large sparse symmetric positive definite system

$$A\mathbf{x} = \mathbf{b} \tag{1}$$

has been a very successful method. If A is nonsymmetric the CG method can not be used for (1) but there have been a number of extensions of the CG method to nonsymmetric problems. One of the earliest of these, and still one of the most used, is the GMRES method of Saad and Schultz[9]. More recent methods include BICGSTAB[14] and CGS[12].

Whether A is symmetric or nonsymmetric, preconditioning is usually required with any of the above methods. Incomplete LU factorization[6] (incomplete Cholesky for symmetric problems) has been a standard preconditioner on serial machines. But this requires the solution of sparse triangular linear systems, a task not well-suited for parallel machines. The same problem exists for SSOR preconditioning. As a consequence, many attempts have been made to carry out ILU and SSOR preconditioning efficiently on parallel machines. One approach that has been reasonably successful on vector machines has been the wavefront ordering (see [1] and [13]), but this has a strong communication penalty on parallel machines. Another approach has been red/black or multicolor orderings. These give good parallel properties but may seriously degrade the rate of convergence of the conjugate gradient method, as compared with the natural ordering. (See e.g., [4] and [7])

The SOR iteration itself, however, does not suffer this degradation. Indeed, if the coefficient matrix is consistently ordered with property A, the asymptotic rates of convergence of the natural and red/black orderings are identical (Young[15]); moreover, in practice one quite often sees faster convergence in the red/black ordering than in the natural ordering. This suggests the possible use of SOR as a parallel preconditioner. It cannot be

---

[*]The work of the first author has been partially supported by NASA Grant NAG-1-1529. The work of the second author has been partially supported by NASA Grant NAG-1-1112-FDP.

[†]Department of Computer Science, University of Virginia, Charlottesville, VA.

a preconditioner for the conjugate gradient method on symmetric positive definite systems since the corresponding preconditioned matrix is not symmetric. But this restriction does not apply to nonsymmetric systems and conjugate-gradient type methods such as GMRES. In fact, Saad[8] showed promising results using several steps of Gauss-Seidel iteration as a preconditioner in conjunction with the GMRES iteration. Shadid and Tuminaro[10] have also reported experiments using one Gauss-Seidel step as a preconditioner.

2 SOR on a Serial Machine

In an earlier paper[3], the authors showed that several SOR steps gave a promising preconditioner on serial machines. This extended the results of [8] by showing that adding the SOR parameter ω did indeed provide benefit, and the results of [10], by showing that only one step may be rather ineffective. We summarize briefly in this section some of these experimental results from [3], which were carried out on an IBM RS/6000, Model 750. One test problem used in [3] was the simple convection-diffusion equation

$$-(u_{xx} + u_{yy}) + \sigma(u_x + u_y) = f(x,y) \tag{2}$$

on the unit square with Dirichlet boundary conditions and with σ constant; for the numerical results shown here $\sigma = 5$. The equation (2) is discretized by standard five-point finite differences using centered differences for the first derivative terms. The right hand side f of (2) is chosen so that the exact solution of the discrete system is known.

Table 1 gives some results using the algorithm GS(k)-GMRES(m), in which we use m GMRES vectors before restarting and k Gauss-Seidel iterations as the preconditioner. GMRES was implemented with unmodified Gram-Schmidt orthogonalization. The table shows iterations and CPU times in seconds for various values of k and m for the problem size $151^2 = 22,801$ equations. The convergence criterion is $\|r\|_2 < 10^{-6}\|A\|_\infty$ and the initial approximation is zero.

Algorithm	Natural iters	Natural time	Red-Black iters	Red-Black time
GMRES(5)	4720	541	4720	548
GS(1)-GMRES(5)	2069	323	1180	170
GS(5)-GMRES(5)	296	99	242	81
GMRES(10)	2455	335	2455	339
GS(1)-GMRES(10)	919	160	659	108
GS(5)-GMRES(10)	157	53	140	47
GS(10)-GMRES(20)	70	40	73	43

TABLE 1
Equation (2). 151^2 equations

Table 1 shows several things: Even one step of Gauss-Seidel is a useful preconditioner for this problem but several Gauss-Seidel steps are much better, corroborating the results of Saad[8]. The use of the red/black ordering has no effect on GMRES itself, as expected, but has a surprisingly beneficial effect when the Gauss-Seidel preconditioner is used. Increasing the number of GMRES vectors from 5 to 10 has a beneficial effect but relatively little effect when increasing from 10 to 20.

Table 1 gives results for only the Gauss-Seidel iteration, but in [3] we also showed that SOR could increase the convergence rate by a factor of 2 or more. As an example, Figure 1 shows the effect on the iteration count of using an SOR relaxation parameter ω.

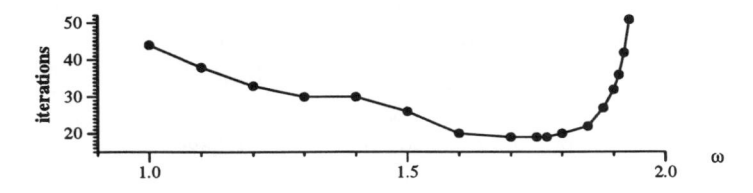

FIG. 1. *Iterations for 2601 Equations, SOR(5)-GMRES(10), Red/Black Ordering*

3 Parallel Results

We next give the results of experiments on two parallel machines, 16 processors of an Intel Paragon and 16 processors of an IBM SP-2. All results are again for the problem (2). We are concerned with parallel efficiency only so we will use a fixed number of iterations with no concern for convergence. Thus, all results will be for 10 iterations of the method SOR(10)-GMRES(10), with $\omega = 1.8$ for SOR.

Table 2 gives speedups on the Paragon for a fixed problem size, 151^2 equations, as used in the serial results of Table 1, as well as a larger problem: 301^2 equations. As expected, Table 2 shows decreasing parallel efficiency as the number of processors increases, and in the sequel we will give only scaled speedup results. For these results, ideally the amount of work would increase exactly in proportion with the increase in the number of processors, but it is difficult to achieve this exact increase by adjusting the grid size. Table 3 shows the problem sizes used and the "work growth factors". For example, for the "small" problem, in which 151^2 is the single processor problem, for 4 processors the grid is 303×303 and the amount of work is 4.04 times the work for the single processor problem, as close as we could get to 4. The work is measured by floating point operations.

	151^2 Eq.		301^2 Eq.	
P	time	speedup	time	speedup
1	4.90	-	18.17	-
2	2.80	1.75	9.65	1.88
4	1.69	2.90	5.15	3.53
8	1.19	4.11	2.95	6.17
16	0.88	5.56	1.79	10.12

TABLE 2
Paragon Speedups, SOR(10) - GMRES(10)

Table 4 shows the total time (T), Mflops (M), scaled speedups (S) and parallel efficiencies (E) for the problem sizes of Table 3. The efficiency is obtained by dividing the speedup by the number of processors. The parallel efficiencies shown in Table 4 are quite good but not perfect. To better understand where efficiency is being lost, we show in Table 5 the times, speedups and efficiencies separately for each of the major components of the algorithm: the SOR preconditioning, the matrix-vector multiplies, the Gram-Schmidt orthogonalization,

P	small		large	
1	151	-	301	-
2	215	2.03	427	2.02
4	303	4.04	603	4.02
8	429	8.10	853	8.04
16	607	16.22	1207	16.11

TABLE 3
Problem Sizes and Work Growth Factors

	small				large			
P	T	M	S	E	T	M	S	E
1	4.9	7.9	-	-	18.2	8.5	-	-
2	5.1	15.4	1.9	0.96	18.3	17.0	2.0	0.99
4	5.2	30.1	3.8	0.94	18.6	33.5	3.9	0.98
8	5.2	60.4	7.5	0.94	18.3	67.9	7.9	0.99
16	5.3	118.5	14.7	0.92	18.5	134.9	15.7	0.98
	10 iterations of SOR(10)-GMRES(10)							

TABLE 4
Paragon Times (T), Mflops (M), Scaled Speedup (S), Efficiency (E).

and everything else (Miscellaneous). The first five rows of Table 5 are for the small problem of Table 4 and the last five rows for the large problem.

	SOR			Matrix-Vector			Gram-Schmidt			Miscellaneous		
P	T	S	E	T	S	E	T	S	E	T	S	E
1	4.13	-	-	0.34	-	-	0.18	-	-	0.22	-	-
2	4.20	2.0	0.98	0.34	2.0	1.00	0.34	1.1	0.53	0.25	1.7	0.87
4	4.25	3.9	0.97	0.34	4.0	0.99	0.47	1.5	0.39	0.31	2.8	0.71
8	4.15	8.0	1.00	0.34	8.0	1.00	0.51	2.9	0.36	0.33	5.3	0.66
16	4.22	15.7	0.98	0.34	16.0	1.00	0.54	5.3	0.33	0.36	9.6	0.60
1	15.39	-	-	1.28	-	-	0.71	-	-	0.75	-	-
2	15.04	2.0	1.02	1.30	2.0	0.99	1.24	1.2	0.58	0.81	1.9	0.93
4	15.24	4.0	1.01	1.30	3.9	0.99	1.40	2.0	0.51	0.84	3.6	0.90
8	14.95	8.2	1.03	1.30	7.9	0.98	1.47	3.9	0.49	0.88	6.8	0.85
16	15.06	16.3	1.02	1.29	15.8	0.99	1.56	7.3	0.46	0.92	13.2	0.82

TABLE 5
Paragon Times (T), Speedup (S), Efficiency (E) for Different Parts

Table 5 shows that the SOR and matrix-vector multiplication parts of the algorithm have almost perfect scaled speedup; good speedups for these parts are of course to be expected. Indeed, SOR shows super-linear speedup for the large problem. This is due to the distribution of the problem across the processors. The bulk of the computation in SOR is done in a pair of nested loops that sweep over the interior of each processor's set of grid points. As the problem size grows the local grid becomes more rectangular; then both inner loops become larger and the resulting "vectors" are longer, so SOR executes a little faster.

We timed execution of the algorithm on each processor and the results given in Table 4 are the largest of the individual processor times. We timed the components the same way. As a result the sum of the component times in Table 5 is slightly greater than the total time given in Table 4; for example, the total of the component times for 16 processors and the large problem is 18.83, as compared with the actual total time of 18.5 from Table 3.

The Miscellaneous part also has fairly good speedup. This part consists of the Hessenberg matrix factorization and solve, the final iterate update, some vector add, scale and set operations, and an inner product. The Hessenberg matrix factorization and solve are done in serial mode, replicated on every processor, but the times for these are so small that they have little impact on the speedup of the algorithm. The least efficient part of the algorithm is the Gram-Schmidt orthogonalization, since it contains several inner products.

It is important to note that proper memory management is critical for obtaining the good speedups of Tables 4 and 5. In order to make our code portable we initially implemented the algorithm using *if77*, the Intel Scalable Systems FORTRAN 77, without any of the VAX/VMS FORTRAN, IBM/VS, or MIL-STD-1753 enhancements, particularly POINTER variables or ALLOCATE/DEALLOCATE statements. Instead we used static memory allocation, and this caused a great deal of paging by the operating system. As a result, stores into the data structure that holds the GMRES vectors took about 30% of the total runtime. By using dynamic memory allocation with the ALLOCATE/DEALLOCATE statements, we reduced the cost of doing these stores by about a factor of eight.

Tables 6 and 7 show corresponding results on an IBM SP-2. Here the times are considerably smaller than those for the Paragon, which is consistent with the relative speeds of the processors on the two machines. But the speedups and parallel efficiencies, although not bad, are not as good as those for the Paragon. Table 7 does show, however, that the parallel efficiency of the Gram-Schmidt part of the algorithm is much better on the SP-2.

Some of the results in Tables 5 and 7 are a little rough. We believe this is due to interrupts and contention for shared resources, as discussed in [2] and [11].

	small				large			
P	T	M	S	E	T	M	S	E
1	0.8	46.8	-	-	3.2	47.8	-	-
2	0.9	88.6	1.9	0.93	3.3	95.3	2.0	0.99
4	0.9	176.8	3.7	0.94	3.4	182.0	3.8	0.95
8	1.0	306.7	6.5	0.81	3.4	369.1	7.7	0.96
16	0.9	672.3	14.2	0.89	3.6	701.7	14.6	0.91
10 iterations of SOR(10)-GMRES(10)								

TABLE 6

SP-2 Times (T), Mflops (M), Speedup (S), Efficiency (E)

4 Summary and Conclusions

In [3], we showed that several SOR iterations could be a good preconditioner for GMRES, at least on certain problems. The preliminary results of the present paper show that the SOR-GMRES algorithm has potentially good parallel properties. Additional experiments with larger numbers of processors and other problems will be necessary to establish more

	SOR			Matrix-Vector			Gram-Schmidt			Miscellaneous		
P	T	S	E	T	S	E	T	S	E	T	S	E
1	0.68	-	-	0.05	-	-	0.08	-	-	0.01	-	-
2	0.73	1.9	0.93	0.06	1.8	0.91	0.10	1.7	0.84	0.02	1.6	0.81
4	0.72	3.8	0.94	0.06	3.6	0.91	0.11	3.0	0.75	0.02	3.3	0.82
8	0.85	6.4	0.80	0.08	4.9	0.62	0.18	3.7	0.46	0.02	6.3	0.79
16	0.76	14.3	0.89	0.07	11.7	0.73	0.14	9.9	0.62	0.02	12.0	0.75
1	2.65	-	-	0.19	-	-	0.34	-	-	0.06	-	-
2	2.66	2.0	0.99	0.22	1.7	0.87	0.35	1.9	0.95	0.07	1.9	0.93
4	2.81	3.8	0.94	0.20	3.9	0.97	0.36	3.8	0.94	0.07	3.8	0.95
8	2.75	7.7	0.96	0.21	7.1	0.89	0.40	6.8	0.85	0.07	7.0	0.87
16	2.92	14.5	0.91	0.21	14.6	0.91	0.43	12.7	0.79	0.07	14.8	0.92

TABLE 7
SP-2 Times (T), Speedup (S), Efficiency (E) for Different Parts

definitively the usefulness of SOR as a parallel preconditioner.

References

[1] C. Ashcraft and R. Grimes. On Vectorizing Incomplete Factorizations and SSOR Preconditioners. *SIAM J. Sci Stat. Comput.*, 9:122–151, 1988.

[2] S. Bokhari. Communication Overhead on the Intel Paragon, IBM SP-2, & Meiko CS-2. Technical Report Int28, ICASE, 1995.

[3] M. DeLong and J. Ortega. SOR as a Preconditioner. *Applied Numerical Mathematics*, 18:431–440, 1995.

[4] I. Duff and G. Meurant. The Effects of Ordering on Preconditioned Conjugate Gradients. *BIT*, 29:635–657, 1989.

[5] G. Golub and C. van Loan. *Matrix Computations*. The Johns Hopkins University Press, Baltimore, MD, 1983.

[6] J. Meijernick and H. van der Vorst. An Iterative Solution Method for Linear Systems of Which the Coefficient Matrix is a Symmetric M-Matrix. *Mathematics of Computation*, 31(137):148–162, 1977.

[7] J. Ortega. Orderings for Conjugate Gradient Preconditionings. *SIAM J. Optimization*, 1(4):565–582, 1991.

[8] Y. Saad. Highly Parallel Preconditioners for General Sparse Matrices. In G. Golub, M. Luskin, and A. Greenbaum, editors, *Recent Advances in Iterative Methods*, pages 165–199. Springer-Verlag, 1994.

[9] Y. Saad and M. Schultz. GMRES: A Generalized Minimal Residual Algorithm for Solving Nonsymmetric Linear Systems. *SIAM J. Sci. Stat. Comput.*, 7(3):856–869, 1986.

[10] J. Shadid and R. Tuminaro. A Comparison of Preconditioned Krylov Methods on a Large-Scale MIMD Machine. *SIAM J. Sci. Stat. Comput.*, 15:440–459, 1994.

[11] H. Shirley, W. Reynolds, and S. Seidel. Communication on the Intel Paragon. Technical Report CS-TR-95-07, Michigan Technological University, 1995.

[12] P. Sonneveld. CGS, a Fast Lanczos-Type Solver for Nonsymmetric Linear Systems. *SIAM J. Sci. Stat. Comput.*, 10(1):36–52, 1989.

[13] H. van der Vorst. High Performance Preconditioning. *SIAM J. Sci. Stat. Comput.*, 10:1174–1185, 1989.

[14] H. van der Vorst. Bi-CGSTAB: a Fast and Smoothly Converging Variant of BI-CG for the Solution of Nonsymmetric Linear Systems. *SIAM J. Sci. Stat. Comput.*, 13(2):631–644, 1992.

[15] D. Young. *Iterative Solution of Large Linear Systems*. Academic Press, New York, 1971.

Chapter 13
A View of Conjugate Gradient-Related Algorithms for Nonlinear Optimization

J.L. Nazareth*

Abstract

This article gives a broad overview of past developments and future trends in research into conjugate gradient-related algorithms for minimizing a high-dimensional nonlinear function.

Keywords: conjugate gradients, nonlinear CG, limited memory algorithms, reduced Hessian QN, successive affine reduction, variable storage algorithms

1 Introduction

The purpose of this article is to provide some background and perspective on past developments and future research trends in the area of conjugate gradient-related algorithms for the nonlinear unconstrained optimization problem:

(1) $$\text{minimize } f(x), \quad x \in R^n,$$

where $f : R^n \to R$ is smooth and n is usually large. The article also provides a context for other articles in this volume that address specific nonlinear CG-related topics.

As an aid to the reader, common acronyms for algorithms encountered in this article are gathered together into a glossary. This is listed in an appendix, to which the reader can refer as the need arises.

2 Standard Nonlinear CG Algorithms

The conjugate gradient method for nonlinear minimization, which was developed in the early 1960's by Fletcher and Reeves [13], coupled a variant of the fundamental *linear CG-direction* of Hestenes and Stiefel [19] with the *line search* technique of Davidon [8], thereby yielding an algorithm suitable for minimizing nonlinear functions. (The name "Reeves" is not well known to the optimization community apart from his connection with this algorithm, and thus it may be worth mentioning that he was Fletcher's graduate research advisor at the Electronic Computing Laboratory, University of Leeds.)

---

*Department of Pure and Applied Mathematics, Washington State University, Pullman, WA 99164-3113 and Department of Applied Mathematics, University of Washington, Seattle, WA 98195. E-mail: nazareth@amath.washington.edu.

An iteration of the Fletcher-Reeves (FR) algorithm was similar to that of the variable metric algorithm of Davidon [8], in the reformulation of Fletcher and Powell [12] popularly called the DFP algorithm. However, the FR algorithm only required $O(n)$ storage, in contrast to DFP, which required $O(n^2)$. The former was therefore more readily applicable to nonlinear minimization problems of high dimension. The basic algorithm is stated in [13] as follows:

x_0 = arbitrary
g_0 = gradient of f at x_0, $d_0 = -g_0$
x_{k+1} = position of minimum of f on line through x_k in direction d_k
g_{k+1} = gradient of f at x_{k+1}
$\beta_k = \|g_{k+1}\|^2 / \|g_k\|^2$
$d_{k+1} = -g_{k+1} + \beta_k d_k$.

Henceforth, we shall also use the notation $y_k = g_{k+1} - g_k$ for the gradient change that corresponds to the step $s_k = x_{k+1} - x_k$. An *implementable* version of the FR algorithm was tested on some low-dimensional problems and shown to be viable, as reported in [13].

Hestenes and Stiefel explicitly describe conjugate gradients in their 1952 paper as 'an iterative method that terminates' (see O'Leary [40]). Nevertheless, their algorithm increasingly came to be viewed by numerical analysts as a member of the family of *direct* methods for solving a positive definite system of linear equations or, equivalently, for minimizing an associated strictly convex quadratic function. Thus, although the Fletcher-Reeves algorithm was a straightforward generalization of the original linear CG approach, it is also important to recognize that it represented a significant departure from the prevalent view of conjugate gradients in the 1960's, and that it led to a fresh and novel approach to minimizing nonlinear functions. (This combination of obviousness and novelty is often characteristic of a good idea.) Later, the work of Reid [48] in the early 1970's changed the perception of *linear* conjugate gradients back from 'direct' to 'iterative', an evolution that was in accord with nonlinear CG usage at the time. This back and forth cross-fertilization, and evolutionary growth in understanding of conjugate gradients, highlights the synergistic connections between the linear and nonlinear facets of the subject.

A key variant of the Fletcher-Reeves algorithm was developed by Polak and Ribière [45] and Polyak [46]. Here the quantity β_k was redefined to be $\beta_k = y_k^T g_{k+1} / g_k^T g_k$. The Polak-Ribière (PR) refinement proved to be important in the nonlinear setting. The third main choice for β_k is the original Hestenes-Stiefel (HS) choice, namely, $\beta_k = y_k^T g_{k+1} / y_k^T d_k$. The three forms are equivalent in the setting of minimizing a positive definite quadratic form using exact line searches, i.e., in the linear CG setting with infinite-precision arithmetic. Additionally, the HS and PR forms are identical on arbitrary nonlinear functions, provided that line searches are exact.

For an excellent survey of what is known about the relative merits of the three main variants (HS, FR and PR) in the nonlinear CG setting, from both convergence analysis and practical standpoints, see Section 4 of Nocedal [39]. Section 3 of the article by Nocedal in this volume, by way of introduction, also gives the correspondences between standard linear and nonlinear CG algorithms.

The Fletcher-Reeves approach was rooted in the notion of conjugate directions. A large number of other CG variants have been proposed from this point of view, which seek to retain conjugacy when the initial direction is not along the negative gradient (Beale [1]) and/or when line searches are not exact (e.g., Dixon [10]). For instance, the three-term-recurrence (TTR) algorithm, Nazareth [27], Dixon et al. [11], which is closely connected to the Lanczos process in the linear setting, is able to simultaneously relax *both* requirements. The overall iteration is similar to the foregoing FR algorithm, but with the search direction replaced by

$$d_{k+1} = -y_k + \beta_k d_k + \alpha_k d_{k-1} \text{ where } \beta_k = y_k^T y_k / y_k^T d_k, \quad \alpha_k = y_{k-1}^T y_k / y_{k-1}^T d_{k-1},$$

and with the initial search direction chosen to be an arbitrary direction of descent. If this initial direction is along the negative gradient and line searches are exact, then the TTR, FR, HS and PR directions are all parallel to one another on a positive definite quadratic. A drawback of the TTR algorithm is that it does not guarantee a descent direction on more general functions, i.e., that $g_{k+1}^T d_{k+1} < 0$ holds, although in practice the direction almost always satisfies this condition (see also Section 5.2).

Many other references to CG variants can be found in Nazareth [33]. Suffice it to say that none of these variants has proved to be significantly more effective than the standard CG approach (so far).

A related approach, which coincidentally was published in the *same* year as the Fletcher-Reeves paper, is the PARTAN algorithm of Shah et al. [49]. On a quadratic function, PARTAN produces identical iterates to the FR algorithm (see Luenberger [23]), and it therefore has the finite termination property (meaning that whenever line searches are exact and infinite-precision arithmetic is used, the algorithm terminates in a finite number of steps, usually $\leq n$). PARTAN is interesting because its basic cycle is very similar to the algorithm of Nesterov [36] for which an important global efficiency result on strongly convex functions has been established. Whether or not Nesterov's algorithm has finite termination on quadratics and the precise nature of the connection between the two algorithms are relatively simple and worthwhile topics for further investigation.

3 VSCG and L-QN Methods

3.1 Beginnings

The germ for a new line of development can be found in two reports, Nazareth [25] and Perry [42], that appeared during the first quarter of 1976. They explored key connections between conjugate gradient and quasi-Newton algorithms and initiated the process that was to result in nonlinear CG-related methods becoming much less reliant on notions of conjugacy.

The report [25], which viewed the subject through a conjugate-gradients lens, explored a key *structural relationship* between search vectors developed by the CG and BFGS algorithms on convex quadratics and on arbitrary smooth functions, and used it to derive a family of variable storage conjugate gradient algorithms (VSCG) whose search directions were preconditioned using a variable metric (defined *implicitly* from stored vectors). Concurrently, the method was implemented and tested as part of the optimization project described in [26]. The BFGS-CG relationship, the resulting VSCG algorithms and a sample of numerical results eventually appeared in [28].

The report [42], which viewed the subject through a quasi-Newton lens, replaced the conjugacy relation $d_{k+1}^T y_k = 0$ by a relation motivated by QN updates, namely, $d_{k+1}^T y_k = -g_{k+1}^T s_k$. The resulting search direction is as follows:

(2) $$d_{k+1} = -g_{k+1} + \beta_k d_k, \text{ where } \beta_k = (y_k - s_k)^T g_{k+1} / y_k^T d_k.$$

Perry [42] formulated and tested a modified conjugate gradient algorithm (PMCG) that used this search direction. When line searches are exact then $s_k^T g_{k+1} = 0$, and the foregoing

PMCG direction is identical to the HS or PR directions on arbitrary functions. (Note also that the usual line search exit condition $y_k^T s_k > 0$ does not guarantee that the PMCG direction is a direction of descent.) The results of the investigation of the PMCG algorithm eventually appeared in [44].

The foregoing dichotomy of approach, namely, orientation towards the CG or QN aspects of the subject, has persisted through subsequent developments in the CG-related field, as will be seen again in Section 4.

3.2 Developments

The work on the BFGS-CG relationship was advanced in an important way by Buckley [2], who set out, as stated in the introduction of this paper, "to extend this relationship and to show that it is specific to the BFGS update". Referring to [3], Buckley continues: "In a companion paper the author has discussed how the resulting theorem may be used to motivate a new conjugate gradient type of minimization algorithm". Other related results along these lines can also be found in Nocedal [37] and in Nazareth and Nocedal [34], [35]. The approach was further extended and studied in detail by Buckley and LeNir [5], and made available as usable mathematical software, Buckley [4].

Perry's PMCG approach was extended by Shanno [50]. The latter's CONMIN algorithm combined three main ingredients: a) a modified PMCG search direction, motivated by a 'symmetrization' technique devised by Dennis and Schnabel [9] (in contrast to PMCG, the weak condition $y_k^T s_k > 0$ ensures that Shanno's direction is a direction of descent), b) use of Beale-type restarts, see Powell [47], and c) use of Oren-Luenberger-Spedicato-type scaling [41]. The symmetrization motivation is now defunct, because the direction can be motivated much more directly and naturally from the BFGS-CG relation. Restarts enhance efficiency in CONMIN, as in the standard CG algorithm, but their use in a more general limited-memory setting is open to question (see the discussion in Perry [43] and Nocedal [39]). Scaling, however, has proved to be a *key* ingredient in a limited memory setting.

The two-step variable metric method (TSVM) of Perry [43] was, in turn, a refinement of CONMIN. It uses a search direction formed from BFGS updates over the *two* most recent steps (expressed implicitly using vectors) and also incorporates scaling into the algorithm.

3.3 Limited Memory - BFGS

By sifting through the foregoing developments, and in particular, by recognizing the key role played by a certain *product/additive representation* of the inverse BFGS update, which permitted continuous updating of the inverse Hessian approximation, say H_{k+1}, to be performed efficiently in a limited-memory setting, Nocedal [38] found *the right synthesis and achieved a breakthrough* with the L-BFGS algorithm. This representation takes the following form:

$$(3) \qquad H_{k+1} = \Omega_k H_k \Omega_k^T + \frac{s_k s_k^T}{y_k^T s_k}, \text{ and } \Omega_k = \left(I - \frac{s_k y_k^T}{y_k^T s_k}\right),$$

where (s_k, y_k) denotes the current step and gradient-change respectively, and the quantity H_k is defined *recursively* in an analogous way, using the pair (s_{k-1}, y_{k-1}); the recursion can be terminated after going back *any* desired number of steps, say, m, i.e., at the iterate x_{k-m+1} with its associated pair (s_{k-m+1}, y_{k-m+1}). At this iterate, some simple initializing matrix, for example, a multiple γ_k of the identity matrix, is used in place of an update derived from information at a prior iterate. An appropriate choice, for example,

$\gamma_k = y_k^T s_k / y_k^T y_k$ motivated by CONMIN, is very important in practice. The steps and gradient changes themselves are stored and no matrices are formed explicitly, and the search direction $-H_{k+1}g_{k+1}$ is obtained efficiently using a suitable recurrence relation. (In the absence of these enhancements, a cycle of the algorithm would be analogous to a cycle of the algorithm stated in Section 2, with the last two items replaced by the update of H_k and the foregoing definition of search direction, respectively.) The approach generalizes Perry's TSVM [43], and it is studied in detail by Liu and Nocedal [21] and Gilbert and Lemarechal [15]. More recently, useful new limited-memory representations were discovered by Byrd, Nocedal and Schnabel [7].

Let us temporarily set aside the issue of implicit *representation* of Hessian approximations and efficient *generation* of search directions, and instead focus on a conceptual L-BFGS algorithm in which storage is not at a premium, i.e., let H_{k+1} be given *explicitly* by an $n \times n$ matrix. Then the only difference between L-BFGS and the BFGS algorithm is that the former discards all information between the starting iterate x_1 and the iterate x_{k-m+1}, and the latter does not. At first sight, this might appear disadvantageous to L-BFGS, even counter-intuitive when m is very small relative to n. However, if a variable metric algorithm is started far from a solution, and if reasonably rapid progress is made at each iteration, then steps and corresponding gradient changes used to define the metric can rapidly become out of date; it may be preferable simply to begin afresh at each iterate with a properly scaled diagonal matrix and update it over only a few recent steps[1]. Along these lines, Gilbert and Lemaréchal [15], in a valuable contribution on variable storage and limited memory algorithms (they use the two terms interchangeably), observe that "when the available memory increases, the performance of a VS method does not improve much. Roughly speaking, they improve till a fairly small number of updates (say smaller than 20, apparently not depending on the number of variables); beyond that value, they stagnate, or even deteriorate". And Siegel [51] states more emphatically: "except for small n neither for Algorithm 6 nor for Nocedal's method can a significant improvement[2] be obtained by increasing m beyond 5, which, for Nocedal's method, is not a new observation (see e.g. Nocedal [1992])". 'Algorithm 6' in this quotation refers to one of the algorithms developed in [51] and the foregoing 1992 reference is to [39].

The following observations may also help to shed further light on why the L-BFGS algorithm is often surprisingly effective in practice:

1. In the BFGS setting, two successive (inverse) Hessian approximations, say W_{k+1} and W_k, differ[3] by a matrix of rank 2 at most. In contrast, the matrices H_{k+1} and H_k obtained in the conceptual L-BFGS setting will generally differ by a matrix of rank more than 2. Higher-rank changes to the matrices defining the variable metric could well be an advantage.

2. Make the simplifying assumption that $m = n$, and define square matrices $S_m = [s_{k-m+1}, \ldots, s_k]$ and $Y_m = [y_{k-m+1}, \ldots, y_k]$, where s_j denotes a step and y_j the corresponding change in gradient, for $j = k - m + 1, \ldots, k$. It is not unreasonable to assert that the 'ideal' (inverse) Hessian approximation and associated metric, derived from this set of information, is a symmetric matrix, say $H_{k+1} \equiv R_{k+1}^T R_{k+1}$, where

---

[1] Indeed, there is hearsay evidence that L-BFGS can occasionally outperform the BFGS algorithm on *small* dimensional problems.

[2] As long as m is not allowed to approach n, when the latter is not small—see Table 5.7 of Leonard [22].

[3] It is inexpensive to obtain W_{k+1} from the preceding matrix W_k, and in the early days of computing, this sort of efficiency gain was crucial.

R_{k+1} is an $n \times n$ matrix, typically upper triangular, that 'minimizes' the quantity

$$\|R_{k+1}^T R_{k+1} Y_m - S_m\| \tag{4}$$

in some suitable matrix norm. (If desired, this norm could even attach greater weight to more recent steps.) The L-BFGS update can be viewed as an *efficient, finite* procedure for obtaining a positive definite symmetric matrix that *approximately* solves this problem.

In fact, much more can be said along these lines by broadening the discussion to permit $m < n$; and by considering analogues of the foregoing variational principle with Hessian approximations in place of inverse Hessian approximations. The well-developed techniques of the matrix differential calculus are very useful in this regard. However, this topic will not be pursued here.

3. It is often easier to prove convergence in the L-QN setting, because it is simpler to obtain bounds on the condition number of the Hessian approximation when it is defined by a sequence of updates over a bounded number of steps, say m. When $m < n$ the rate of convergence of L-BFGS is, in general, only linear. An interesting theoretical question is whether superlinear asymptotic convergence is achievable when $m = n$ (it is important to note that L-BFGS is not equivalent to the BFGS algorithm even with this assumption); and an interesting empirical counterpart to this question is whether L-BFGS (with $m = n$ and an appropriate choice of scaling) is generally comparable in performance to standard BFGS.

3.4 Summary

As an aid to understanding, it is useful to consider the finite termination properties of VSCG and L-QN under the usual assumptions of strictly convex quadratics and exact line searches. For example, consider the VSCG algorithm formulated in [28], with continuous updating of the metric using a limited-memory SR1 update; the latter can be represented in the form given in Byrd, Nocedal and Schnabel [7], omitting updates that are not positive definite. This algorithm would terminate on quadratics in at most n steps, even when some SR1 updates are omitted. A similar result holds when *any* other limited-memory update from Broyden's positive definite class is substituted for the SR1. (SR1 has the advantage that it requires the least amount of storage per update.) In contrast, consider the corresponding L-SR1 algorithm formulated analogously to L-BFGS, but substituting the limited-memory SR1 update of [7], and again omitting updates that are not positive definite. This does *not* have quadratic termination in general. Similar remarks apply to other L-QN algorithms. Only L-BFGS has the finite termination property.

VSCG algorithms thus have stronger quadratic termination properties than L-QN algorithms. However, unlike the L-BFGS algorithm or indeed any L-QN algorithm based on the positive definite Broyden family, VSCG algorithms share a common disadvantage with standard nonlinear CG algorithms: more stringent requirements than the usual (weak) Wolfe conditions must be imposed on the line search in order to ensure that the search direction is a direction of descent.

As we have seen, VSCG algorithms helped set the stage for limited-memory BFGS algorithms, but the former are not considered to be computationally competitive today. The evolutionary line of development discussed in this section has thus broken loose from its initial moorings in CG methods and has entered the mainstream of variable-metric/quasi-Newton techniques. Limited-memory QN algorithms, a refinement of QN algorithms designed to handle large-scale minimization problems, have become part of the 'technology' of the QN method and have led to practical implementations and quality mathematical software. The most useful of these is L-BFGS.

The limited memory algorithms discussed so far have been considered in a *metric-based line search* setting (for terminology and background, see [33]). L-QN algorithms in *trust region* or *model-based line-search* settings are a current topic of research and may lead to useful implementations in the future (Burke [6]).

Finally, we may observe that a synthesis of the L-SR1 and L-BFGS approaches is possible: a sequence of L-SR1 updates over a set of prior steps (represented as in [7] and safeguarded to ensure positive definiteness) can be concluded with a *single* L-BFGS update over the most recent pair (s_k, y_k). Given a fixed amount of available storage, this hybrid algorithm would permit more updates to be saved than would be possible using L-BFGS over all steps. With exact line searches, it would preserve the finite termination property on quadratics, and it would guarantee descent directions on arbitrary functions under standard restrictions on the line search.

4 RH and SAR Methods

4.1 Beginnings

A fresh line of development began with the *projected Hessian* approach of Fenelon [14] at the beginning of the 1980's. A second approach, termed *successive affine reduction* (SAR), is described in a sequence of articles, Nazareth [29], [30], [31]. A third approach, from yet another point of view, is given by Siegel [51]. These three approaches are related, but were developed independently of one another at well-spaced intervals, and consequently each has a different orientation and character, as even a cursory reading of the cited articles will reveal. (This is in marked contrast to the algorithms of Section 3, where there has been a distinct pattern of evolutionary growth.) A good overview of the underlying ideas and their interconnections, along with useful new contributions, can be found in the recent work of Leonard [22].

4.2 Reduced Hessian Quasi-Newton

Fenelon [14] and Siegel [51] view the subject through a QN lens, and the following key lemmas, which are quoted from Leonard [22], provide the basis for the algorithms. These lemmas use the notation $G_{k+1} = \text{span}\{g_1, \ldots, g_{k+1}\}$ and $D_{k+1} = \text{span}\{d_1, \ldots, d_{k+1}\}$. Also, following Leonard [22], we will subsequently use the term *reduced Hessian* (RH) quasi-Newton for methods of this type.

Lemma 1 (Fenelon): If the BFGS method is used to solve the unconstrained minimization problem with initial Hessian approximation σI ($\sigma > 0$), then $d_{k+1} \in G_{k+1}$ for all k.

Lemma 2 (Siegel): If a variable metric algorithm based on any member of the positive definite Broyden family is used to solve the unconstrained minimization problem with initial Hessian approximation σI ($\sigma > 0$), then $d_{k+1} \in G_{k+1}$ for all k. Moreover, $G_{k+1} = D_{k+1}$, and if $z \in G_{k+1}$ and $w \in G_{k+1}^{\perp}$ then $M_{k+1} z \in G_{k+1}$, $W_{k+1} z \in G_{k+1}$, $M_{k+1} w = \sigma w$ and $W_{k+1} \in w = \sigma^{-1} w$, where M_{k+1} denotes a positive definite symmetric approximation to the Hessian matrix, and $W_{k+1} \equiv M_{k+1}^{-1}$.

The reduced Hessian QN approach (denoted by RH-G in [22]) defines an orthogonal basis for the subspace spanned by G_{k+1}, uses it to form reduced approximate Hessians or their inverses or Cholesky factors (whose dimension $m \times m$ corresponds to that of the subspace and may be considerably smaller than $n \times n$), and derives reduced Hessian or reduced inverse Hessian quasi-Newton algorithms that produce *identical* iterates in exact arithmetic to their corresponding quasi-Newton counterparts. Reduced Hessian *limited-memory* quasi-Newton methods (RH-L-G) are obtained by restricting the subspace G_{k+1} to only a few recent gradients. Variants are obtained by using subspaces spanned by directions in place of gradients (identified by the acronyms RH-P and RH-L-P, where the symbol P comes from the use of the notation p_k instead of d_k for search directions in [22]). In the setting of *arbitrary* smooth functions, minimal or close to minimal use of storage (see also Section 5.4) and exact line searches, the precise connections between these algorithms and standard nonlinear CG algorithms has not been studied in any detail, so far. The special case of quadratics is discussed in Section 4.4.

For further details on RH methods along with the results of numerical experiments, see Leonard [22]. In particular, the use of appropriate scaling strategies is very important in practice, see Table 5.4 in [22].

4.3 Successive Affine Reduction

Nazareth [29], [30] views the subject through a CG lens, and seeks a conceptual underpinning idea for nonlinear conjugate gradients. This relies on the following principle:

SAR Principle (for background and terminology, see [32], [33]): Make metric-based or model-based, Newton or quasi-Newton estimates of curvature, i.e., approximations to the Hessian or its inverse, in an affine subspace normally of *low dimension*, say m, and determined by the *most recent gradient and step*, and zero, one or more earlier gradients and/or steps.

SAR/BFGS algorithms are developed in [29]. SAR/Newton algorithms are described in [30], [52]. When line searches are exact and minimal storage is used, SAR/BFGS is equivalent to a standard CG algorithm on arbitrary smooth functions. If additionally the function is a convex quadratic then SAR/Newton is also equivalent to a standard CG algorithm (and closely related on arbitrary functions). At the other extreme, when $O(n^2)$ storage is employed, SAR/Newton is equivalent to the Newton algorithm, and SAR/BFGS is equivalent to the BFGS algorithm. (As noted in Section 3.3, the latter is not true of the limited-memory BFGS algorithm when the n most recent step and gradient-change pairs are used, although this is not necessarily a disadvantage.) Algorithms based on the SAR principle thus provide a *true* continuum between Newton and quasi-Newton methods on the one hand and a standard nonlinear CG method on the other.

How is the transition from linear convergence (SAR subspace of dimension 2) to superlinear convergence (SAR subspace of dimension n) effected in the SAR/BFGS method? How is the transition from linear convergence (SAR subspace of dimension 2) to quadratic convergence (SAR subspace of dimension n) effected in the SAR/Newton method? These are interesting *theoretical* issues[4], and their analysis may shed light on the formulation of more effective SAR algorithms.

From a *practical* standpoint, numerical evidence in [29] suggests that there are advantages to using affine subspaces of dimension 2, 3 or 4 in the SAR/BFGS algorithm,

---

[4] Use of $O(n^2)$ storage is not practical for high-dimensional problems.

but one quickly reaches the point of diminishing returns. The situation is analogous to experience with L-BFGS. Subspaces of dimension more than 2 have not yet been studied in the SAR/Newton case (see also Section 5.4.1).

As in the L-BFGS algorithm, the use of appropriate scaling strategies is of practical importance. This topic has also not been studied, in any detail, in the SAR setting.

4.4 Summary

Although arrived at independently, the methods developed by Fenelon [14] and Siegel [51], premised on the Lemmas of Section 4.2, have much in common, as shown in Leonard [22]. As with the L-BFGS method, the reduced Hessian algorithms RH-L-G and RH-L-P can be viewed as part of the QN family, refined to handle the case when storage is limited.

The SAR approach is related (clearly RH methods also make estimates of curvature in a subspace), but it is much more general, having both Newton and quasi-Newton algorithmic expressions. By virtue of the underlying CG-oriented perspective, all versions of SAR algorithms with minimal use of storage have a very close connection with standard nonlinear conjugate gradients, as described in Section 4.3.

It is again instructive to examine the finite termination properties and the hereditary properties (satisfying the quasi-Newton relation simultaneously over several steps) of the foregoing algorithms when applied to quadratic functions:

The limited-memory algorithms in Fenelon [14] achieve quadratic termination somewhat unnaturally (in our opinion) by relying on the fact that the reduced Hessian is tridiagonal when a quadratic function is minimized. Their hereditary properties have not yet been studied.

One of the two limited-memory algorithms in Siegel [51], which is related to RH-L-G in [22], does not have the quadratic termination property. The other (called Algorithm 6 in [51] and closely related to RH-L-P in [22]) is shown to be equivalent to the standard CG algorithm when the function is a strictly convex quadratic and line searches are exact; and this result is true, in particular, when minimal or close to minimal storage (see also Section 5.4) is used. The result relies on use of the BFGS update in the algorithm, and it does not hold if another update from the positive definite Broyden family is substituted. The hereditary properties of RH-L algorithms have not yet been studied.

Quadratic termination and hereditary properties of SAR/BFGS have been established in [29]. The use of other updates in the SAR/QN setting has not yet been explored, but some results are easily obtained. For example, if the SAR subspace is of dimension 2, and other updates are used in place of BFGS, it is straightforward to show that the resulting search direction on an arbitrary function is still the standard CG direction, when the line search is exact (see Section 3 (a) of [29]).

In summary, we have again seen the dichotomy, previously encountered in Section 3, between CG and QN-oriented approaches. But, in contrast to the the advanced development of variable storage and limited memory algorithms of Section 3—see also Section 5.3, much remains to be done in order to obtain practical realizations of RH-L and SAR algorithms.

5 Nonlinear CG-Related Algorithms

In Section 2, we described standard nonlinear CG algorithms that are rooted in the idea of conjugate directions. In Sections 3 and 4, we described algorithms that occupy the spectrum (or continuum) between these standard nonlinear CG algorithms and Newton or quasi-Newton algorithms. Let us now take the wheel full circle by considering the 'far end of the continuum', so to speak, where *minimal or close to minimal storage is used*. We will have returned to the arena of CG algorithms, but at a new level of development that is relatively free from reliance on notions of conjugacy.

5.1 Characterization

Without seeking a "definition" in any precise sense of the word, let us characterize a *nonlinear CG-related algorithm* as follows:

- Its storage requirements are *similar* to those of an implemented version of a standard *nonlinear* CG algorithm, i.e., to within a factor of between 1 and, say, 2 or 3. It is generally agreed that a standard CG algorithm benefits, in practice, from the use of a Beale-type restart strategy, see Beale [1] and Powell [47]. Thus we will consider the Beale-type extension as determining the amount of storage required by an implemented standard nonlinear CG algorithm.

- It generates the *same* iterates as a standard *linear* CG algorithm when the function is a strictly convex quadratic, line searches are exact, and the same initialization is used. Thus, in particular, if an algorithm does not have finite termination on a quadratic when line searches are exact, it is not a CG-related algorithm.

We summarize the CG-Related algorithms encountered in the discussion of previous sections, and others that are related, in a table. We then mention some available software in each category, and comment briefly on potential developments.

Category	CG-related Algorithms
Standard CG Variants	FR [13], PR [45], HS [19], Beale [1], PARTAN [49], TTR [27], [11]
Variable-Storage CG Limited-Memory QN	VSGCG [28], BBVSCG [3], [5] PMCG [42], CONMIN [50], TSVM [43], L-BFGS [38], [21], M1QN [15]
Reduced Hessian QN Affine Reduction N Affine Reduction QN	DRecur [14], RH-L-P [51], [22] SAR/MN [30] [52], GPR [20] SAR/BFGS [29], SY [53]

5.2 Standard CG and Variants

Software that implements standard nonlinear CG methods or TTR-based methods includes the following:

- Routine VA14 of the Harwell subroutine library, see Powell [47]. This incorporates the Beale-type restart strategy mentioned earlier.

- Routine OPCG2 from the Numerical Optimization Center library, see Dixon, Ducksbury and Singh [11]. This implements the TTR variant described in Section 2. In contrast to standard CG, the TTR direction does not have to be modified if a starting direction other than the negative gradient is used. In this sense, TTR has its own built-in restart strategy and it has similar storage requirements to the Beale extension of conjugate gradients.

Preconditioning by a fixed or updated diagonal matrix can have a dramatic effect on the performance of a standard nonlinear CG algorithm (see Gill and Murray [17]). Techniques for developing diagonal preconditioners are also given in Gilbert and Lemarechal [15], and they may have a useful role to play in the implementation of nonlinear CG algorithms. If the diagonal preconditioner is just a multiple of the identity matrix, as in the self-scaling approach, then the HS direction remains unchanged and is only altered in length, in contrast to the FR and PR choices. To date, automatic scaling or diagonal preconditioning has not been incorporated *effectively* into standard nonlinear CG software.

5.3 VSCG and L-QN

Software for these methods includes the following:
- Routine BBVSCG in the ACM-TOMS algorithms collection, which implements a VSCG algorithm, see Buckley [4].

- Routine VA15 in the Harwell library, which implements the L-BFGS algorithm, see Liu and Nocedal [21].

- Routine M1QN in the INRIA/MODULOPT library, which also implements a version of the L-BFGS algorithm, see Gilbert and Lemarechal [15].

These implementations are all able to vary the number of updates. As noted earlier in Section 3.4, VSCG algorithms are not considered to be as efficient as L-BFGS.

When minimal storage is employed, the routines conform to our earlier characterization of CG-related methods. The benchmark algorithm in this respect is TSVM, see Perry [43], which corresponds to L-BFGS with the choice $m = 2$, with computations *reformulated* to enhance efficiency and conform more closely to a standard CG framework. Both L-BFGS and TSVM use scaling, but no restarts.

The report of Gill and Murray [17] contains much useful information on the relative performance of *other variants*, and on the use of a more general diagonal preconditioning. New developments based on Byrd, Nocedal and Schnabel [7] (for example, algorithms of the form discussed at the end of section 3.4) could lead to improvements. L-QN based on trust regions is also currently being studied (Burke [6]).

5.4 RH and SAR

There is no user-oriented, fixed-storage or variable-storage software currently available for RH-L or SAR methods (with one exception mentioned below). We shall only discuss potential development of these methods in the CG-related setting of minimal or close to minimal use of storage, i.e. $m = 2, 3$.

5.4.1 Newton Case $m = 2$:
Encouraging results have been obtained on small problems for the SAR/Newton algorithm described in [30], [32], [52]. Here, the second order information is obtained by a few finite differences of function values. This is appropriate when computation of a function value is much cheaper than computation of a gradient vector. A useful implementation of a related approach, arrived at independently from a different point of view, is

- Routine GPR (Generalized Polak-Ribiere) from the Loughborough University of Technology, see Khoda, Liu and Storey [20].

This can be shown to correspond to an SAR/Newton approach with the second-order information obtained using an extra gradient evaluation (and reformulation of the computations to conform more closely to a standard nonlinear CG framework). This alternative is useful when the computation of a gradient vector is not much more than the computation of a few function values, for example, when gradients are obtained by automatic differentation.

Case $m = 3$: This remains to be explored. Suppose a three-dimensional affine subspace is used, which is defined, for example, by the current gradient, the current step, and the previous gradient or step. The Hessian information in this affine subspace, obtained by finite differences, requires about 10 extra function evaluations. Suppose a gradient evaluation is much more costly than a function evaluation — in an extreme case, say it costs n function evaluations, and take $n = 1000$. Then the extra second-order information results in only about a one percent increase in cost over that of obtaining a function value and gradient.

Reformulations analogous to GPR [20], in order to use gradient vectors in place of function values, may also be of interest.

5.4.2 Quasi-Newton Case $m = 2$:
Encouraging numerical supporting evidence for SAR/BFGS is given in [29] for small problems. If the search direction is reformulated for this case, it can be stated as follows:

$$(5) \qquad d_{k+1} = \frac{1}{\theta}\left[-g_{k+1} + \beta_k d_k\right], \text{ where } \beta_k = (y_k - \theta s_k)^T g_{k+1} / y_k^T d_k.$$

The scalar θ is *implicit* in the algorithm, and can be derived a posteriori from the SAR/BFGS direction. Comparing expressions (2) and (5) we see that SAR/BFGS is a refinement of PMCG ($\theta = 1$) with the property that a direction of descent is obtained whenever $y_k^T s_k > 0$.

If another update is substituted for the BFGS it will have an analogous quantity θ associated with it—see also the remark near the end of Section 4.4. Interesting new CG-related algorithms may be obtained by treating θ as an *explicit* parameter of (5), and choosing it to ensure descent via some appropriate variational principle, for example, minimization of the directional derivative at the current iterate.

Another interesting approach, closely related to SAR/BFGS with $m = 2$, is given by Stoer and Yuan [53].

Case $m = 3$: Suppose a three-dimensional affine subspace is used, which is defined by the current gradient, the current step, and the previous gradient or step. For these two choices of subspace, encouraging numerical results have been obtained for SAR/BFGS on small problems. An interesting observation (by Min Zhu), which favours $m = 3$, is that this is the smallest value of m that ensures the quasi-Newton relation is satisfied by the Hessian or Hessian inverse approximation in *both* the full space and the affine reduced space.

For $m = 3$, encouraging results for an RH-L type algorithm are reported in Siegel [51]. See also Leonard [22], Tables 5.8 and 5.9. Stoer and Yuan [53] consider the case $m = 3$ in their SAR-type approach, and their numerical results are also promising.

5.5 Other Approaches

Other useful approaches that would not normally be considered to be nonlinear CG-related include the Truncated Newton (TN) Method, see for example Nash [24], the recently proposed Iterated-Subspace Method (ISM)—see the article by Conn et al. in this volume, and the very recent Discrete Newton Method with Memory—see the article by Nocedal in this volume. These methods use the *linear* conjugate gradient algorithm within an inner iteration in order to obtain a search direction or an improved estimate of the optimal solution. By estimating matrix-vector products required in the inner algorithm by gradient differences, and by using an appropriate Beale-type linear CG algorithm with a starting direction picked up from where the previous inner iteration left off, it *may* be possible to design versions of these methods that also conform to the nonlinear CG-related characterization of Section 5.1 (when minimal or close to minimal storage is employed). However, it is unclear whether such modifications would be helpful or detrimental when used to minimize nonquadratic functions.

6 Conclusions

CG-related algorithms lie at the interface of computational linear equation-solving and computational nonlinear optimization, and they are some of the most aesthetically pleasing algorithms available in these two areas. In Section 4 of [39], Nocedal makes the following important observations concerning nonlinear CG algorithms:

> ... nonlinear conjugate gradient methods possess surprising, and sometimes bizarre properties. The theory developed so far offers fascinating glimpses into their behaviour, but our knowledge remains fragmentary. I view the development of a comprehensive theory of conjugate gradient methods as one of the outstanding challenges in theoretical optimization, and I believe that it will come to fruition in the near future. The theory would not only be a significant mathematical accomplishment, but could result in a superior conjugate gradient method.

And to conclude the same section, he observes:

> Have we failed to discover the "right" implementation of the conjugate gradient method? Is there a simple iteration ... which performs significantly better than all the methods proposed so far, and which has all the desirable convergence properties? Given the huge number of articles proposing new variations of the conjugate gradient method, without much success, the answer would seem to be "no". However, I have always felt the answer is "yes"—but I could say no more.

We have seen in the present overview that there are promising avenues to explore in pursuit of the above goals. Convergence analysis and global efficiency analysis like that of Gilbert and Nocedal [16] and Nesterov [36], respectively, will be needed, along with computational studies to explore numerical performance and relative merit on large-scale problems. It is difficult to predict at the present time whether this research will only confirm the all-too-familiar pattern of erratic performance of nonlinear CG-related algorithms (which would, in turn, imply the need for CG-related software with a *menu* of user-selected options to suit particular problem characteristics) or whether the much sought after "ideal" nonlinear CG formulation will emerge.

7 Acknowledgements

It is a pleasure to thank Loyce Adams, Andy Conn, Mike Leonard and Michael Saunders for helpful comments that improved this article.

References

[1] E.M.L. Beale, *A derivation of conjugate gradients*, in Numerical Methods for Nonlinear Optimization, F.A. Lootsma (Ed.), Academic Press, N.Y., 1972, pp. 39-43.

[2] A. Buckley, *Extending the relationship between the conjugate gradient and BFGS algorithms*, Mathematical Programming, 15 (1978), pp. 343-348.

[3] A. Buckley, *A combined conjugate-gradient quasi-Newton minimization algorithm*, Mathematical Programming, 15 (1978), pp. 200-210.

[4] A. Buckley, *Algorithm 630: BBVSCG - A variable-storage algorithm for function mimimization*, ACM Transactions on Mathematical Software, 11 (1985), pp. 103-119.

[5] A. Buckley and A. LeNir, *QN-like variable storage conjugate gradients*, Mathematical Programming, 27 (1983), pp. 155-175.
[6] J. Burke, private communication, 1995.
[7] R.H. Byrd, J. Nocedal and R.B. Schnabel, *Representations of quasi-Newton matrices and their use in limited memory methods*, Mathematical Programming, 63 (1994), pp. 129-156.
[8] W.C. Davidon, *Variable metric method for minimization*, Argonne National Laboratory, Report ANL-5990 (Rev.), Argonne, Illinois, 1959. (reprinted in SIAM J. Optimization, 1 (1991)).
[9] J.E. Dennis and R.B. Schnabel, *Numerical Methods for Unconstrained Optimization and Nonlinear Equations*, Prentice-Hall, 1983.
[10] L.C.W. Dixon, *Conjugate gradient algorithms: quadratic termination without linear searches*, J.I.M.A., 15 (1975), pp. 9-18.
[11] L.C.W. Dixon, P.G. Ducksbury and P. Singh, *A new three term conjugate gradient method*, Report No. 130, Numerical Optimization Center, The Hatfield Polytechnic, England, 1983.
[12] R. Fletcher and M.J.D. Powell, *A rapidly convergent descent method for minimization*, Computer Journal, 6 (1963), pp. 163-168.
[13] R. Fletcher and C. Reeves, *Function minimization by conjugate gradients*, Computer Journal, 7 (1964), pp. 149-154.
[14] M.C. Fenelon, *Preconditioned Conjugate-Gradient-Type Methods for Large-Scale Unconstrained Optimization*, Ph.D. Dissertation, Stanford University, Stanford, CA, 1981.
[15] J.C. Gilbert and C. Lemaréchal, *Some numerical experiments with variable storage quasi-Newton algorithms*, Mathematical Programming, 45 (1989), pp. 407-436.
[16] J.C. Gilbert and J. Nocedal, *Global convergence properties of conjugate gradient methods of optimization*, Research Report 1268, INRIA, Rocquencourt, France, 1990.
[17] P.E. Gill and W. Murray, *Conjugate gradient methods for large-scale nonlinear optimization*, Tech. Report SOL 79-15, Department of Operations Research, Stanford University, Stanford, CA, 1979.
[18] M.R. Hestenes, *Conjugate Direction Methods in Optimization*, Springer-Verlag (Applications of Mathematics Series, 12), New York and Heidelberg, 1980.
[19] M.R. Hestenes and E.L. Stiefel, *Methods of conjugate gradients for solving linear systems*, J. Res. Nat. Bur. Stds., Section B, 49 (1952), pp. 409-436.
[20] K.M. Khoda, Y. Liu and C. Storey *Optimized software for a generalized Polak-Ribiere algorithm in unconstrained optimization*, Technical Report A156, Department of Mathematical Sciences, Loughborough University of Technology, England, 1992.
[21] D.C. Liu and J. Nocedal, *On the limited memory BFGS method for large-scale optimization*, Mathematical Programming, 45 (1989), pp. 503-528.
[22] M.W. Leonard, *Reduced Hessian Quasi-Newton Methods for Optimization*, Ph.D. Dissertation, University of California, San Diego, CA, 1995.
[23] D.G. Luenberger, *Linear and Nonlinear Programming*, Addison-Wesley (Second Edition), 1984.
[24] S.G. Nash, *User's guide for TN/TNBC: Fortran routines for nonlinear optimization*, Report 397, Mathematical Sciences Department, The Johns Hopkins University, 1984.
[25] J.L. Nazareth, *A relationship between the BFGS and conjugate gradient algorithms*, Tech. Memo. ANL-AMD 282, Argonne National Laboratory, January, 1976. (Presented at the SIAM-SIGNUM Fall 1975 Meeting, San Francisco, CA.)
[26] J.L. Nazareth, *MINKIT - An optimization system*, ANL-AMD Tech. Memo. 305, Argonne National Laboratory, Argonne, IL, 1977.
[27] J.L. Nazareth, *A conjugate direction algorithm for unconstrained minimization without line searches*, J.O.T.A., 23 (1977), pp. 373-387.
[28] J.L. Nazareth, *A relationship between the BFGS and conjugate gradient algorithms and its implications for new algorithms*, SIAM J. Numerical Analysis, 16 (1979), pp. 794-800.
[29] J.L. Nazareth, *The method of successive affine reduction for nonlinear minimization*, Mathematical Programming, 35 (1986), pp. 97-109.
[30] J.L. Nazareth, *An algorithm based upon successive affine reduction and Newton's method*, In: *Proceedings of the Seventh INRIA International Conference on Computing Methods in*

Applied Science and Engineering, (Versailles, France), R. Glowinski and J-L. Lions, Eds., North-Holland, 1986, pp. 641-646.

[31] J.L. Nazareth, *Conjugate gradient algorithms less dependent on conjugacy*, SIAM Review, 28 (1986), pp. 501-511.

[32] J.L. Nazareth, *The Newton and Cauchy perspectives on computational nonlinear optimization*, SIAM Review, 36 (1994), pp. 215-225.

[33] J.L. Nazareth, *The Newton-Cauchy Framework: A Unified Approach to Unconstrained Nonlinear Minimization*, LNCS 769, Springer-Verlag, Berlin, 1994.

[34] J.L. Nazareth and J. Nocedal, *A study of conjugate gradient methods*, Tech. Report SOL 78-29, Department of Operations Research, Stanford University, Stanford, CA, 1978.

[35] J.L. Nazareth and J. Nocedal, *Conjugate direction methods with variable storage*, Mathematical Programming, 23 (1982), pp. 326-340.

[36] Y.E. Nesterov, *A method of solving a convex programming problem with convergence rate $O(1/k^2)$*, Soviet Mathematics Doklady, 27 (1983), pp. 372-376.

[37] J. Nocedal, *On the Method of Conjugate Gradients for Function Minimization*, Ph. D. Dissertation, Rice University, Houston, TX, 1978.

[38] J. Nocedal, *Updating quasi-Newton matrices with limited storage*, Mathematics of Computation, 35 (1980), pp. 773-782.

[39] J. Nocedal, *Theory of algorithms for unconstrained optimization*, Acta Numerica, 1 (1992), pp. 199-242.

[40] D.P. O'Leary, *Conjugate gradients and related KMP algorithms: the beginnings*, in *Linear and Nonlinear Conjugate Gradient-Related Methods*, L. Adams and J.L. Nazareth (Eds.), SIAM, Philadelphia, 1996.

[41] S. Oren and E. Spedicato, *Optimal conditioning of self-scaling variable metric algorithms*, Mathematical Programming, 10 (1976), pp. 70-90.

[42] A. Perry, *A modified conjugate gradient algorithm*, Discussion Paper 229, Center for Mathematical Studies in Economics and Management Science, Northwestern University, Evanston, IL, (March, 1976, revised December, 1976).

[43] A. Perry, *A class of conjugate gradient algorithms with a two-step variable metric memory*, Discussion Paper 269, Center for Mathematical Studies in Economics and Management Science, Northwestern University, Evanston, IL, January, 1977.

[44] A. Perry, *A modified conjugate gradient algorithm*, Operations Research, 26 (1978), pp. 1073-1078.

[45] E. Polak and G. Ribière, *Note sur la convergence de methode de directions conjuguees*, Revue Francaise d'Informatique et de Recherche Operationnelle, 16 (1969), pp. 35-43.

[46] B.T. Polyak, *The conjugate gradient method in extremal problems*, USSR Comp. Math. and Math. Phys., 9 (1969), pp. 94-112.

[47] M.J.D. Powell, *Restart procedures for the conjugate gradient method*, Mathematical Programming, 12 (1977), pp. 241-254.

[48] J.K. Reid, *On the method of conjugate gradients for the solution of large sparse systems of linear equations*, in *Large Sparse Sets of Linear Equations*, Academic Press, New York, 1971, pp. 231-254.

[49] B.V. Shah, R.J. Buehler and O. Kempthorne, *Some algorithms for minimizing a function of several variables*, J. Soc. Ind. & Appl. Math., 12 (1964), pp. 74-91.

[50] D.F. Shanno, *Conjugate gradient methods with inexact searches*, Mathematics of Operations Research, 3 (1978), pp. 244-256.

[51] D. Siegel, *Implementing and modifying Broyden class updates for large scale optimization*, Report DAMTP 1992/NA12, University of Cambridge, England, 1992.

[52] L.E. Stephenson and J.L. Nazareth, *Successive affine reduction and a modified Newton's method*, Tech. Report 95-1, Department of Pure and Applied Mathematics, Washington State University, Pullman, WA, 1995.

[53] J. Stoer and Y. Yuan, *A subspace study on conjugate gradient algorithms*, preprint, 1993.

APPENDIX

Glossary of Common Acronyms

Acronym	Explanation
BFGS	Broyden-Fletcher-Goldfarb-Shanno
CG	Conjugate Gradients
DFP	Davidon-Fletcher-Powell
FR	Fletcher-Reeves
HS	Hestenes-Stiefel
ISM	Iterated-Subspace Method
L-BFGS	Limited Memory BFGS
L-QN	Limited Memory Quasi-Newton
L-SR1	Limited Memory SR1
LM	Limited Memory
MN	Modified Newton
PARTAN	Parallel Tangents
PMCG	Perry's Modified Conjugate Gradients
PR	Polak-Ribière
QN	Quasi-Newton
RH	Reduced Hessian
RH-L	Reduced Hessian with Limited Memory
RH-L-G	RH-L and Gradient-based subspaces
RH-L-P	RH-L and Direction-based subspaces
SAR	Successive Affine Reduction
SR1	Symmetric Rank One
TN	Truncated Newton
TSVM	Two-Step Variable-Metric
TTR	Three Term Recurrence
VS	Variable Storage
VSCG	Variable Storage Conjugate Gradients